T0200679

Civil Engineering PE Practice Exams

Practice Exams

Breadth and Depth

Civil Engineering PE Practice Exams

Breadth and Depth

Indranil Goswami, Ph.D., P.E.

Second Edition

New York   Chicago   San Francisco   Athens   London   Madrid
Mexico City   Milan   New Delhi   Singapore   Sydney   Toronto

Library of Congress Control Number: 2021940043

McGraw Hill books are available at special quantity discounts to use as premiums and sales promotions, or for use in corporate training programs. To contact a representative please visit the Contact Us page at www.mhprofessional.com.

Civil Engineering PE Practice Exams: Breadth and Depth, Second Edition

Copyright ©2021 by McGraw Hill. All rights reserved. Printed in the United States of America. Except as permitted under the United States Copyright Act of 1976, no part of this publication may be reproduced or distributed in any form or by any means, or stored in a data base or retrieval system, without the prior written permission of the publisher.

1 2 3 4 5 6 7 8 9   CCD   26 25 24 23 22 21

ISBN 978-1-260-46692-8
MHID 1-260-46692-2

The pages within this book were printed on acid-free paper.

Sponsoring Editor
Ania Levinson

Editorial Supervisor
Donna M. Martone

Production Supervisor
Pamela A. Pelton

Acquisitions Coordinator
Elizabeth M. Houde

Project Manager
Warishree Pant,
KnowledgeWorks Global Ltd.

Copy Editor
Girish Sharma,
KnowledgeWorks Global Ltd.

Proofreader
Suvairiyyath Beevi,
KnowledgeWorks Global Ltd.

Art Director, Cover
Jeff Weeks

Composition
KnowledgeWorks Global Ltd.

Information contained in this work has been obtained by McGraw Hill from sources believed to be reliable. However, neither McGraw Hill nor its authors guarantee the accuracy or completeness of any information published herein, and neither McGraw Hill nor its authors shall be responsible for any errors, omissions, or damages arising out of use of this information. This work is published with the understanding that McGraw Hill and its authors are supplying information but are not attempting to render engineering or other professional services. If such services are required, the assistance of an appropriate professional should be sought.

This edition is dedicated to Lali and Ishan, who tolerate me every day and give me their continuous love and support.

—*Indranil*

ABOUT THE AUTHOR

Dr. Indranil Goswami has a Ph.D. in Civil Engineering from Johns Hopkins University in Baltimore, Maryland. He has taught at Morgan State University and served as the department's graduate coordinator. He currently works as a consulting engineer in Baltimore, Maryland. Dr. Goswami is a registered Professional Engineer in the State of Maryland and he has served as President of the Baltimore Chapter of the Maryland Society of Professional Engineers. He taught PE (civil) and FE review courses in Baltimore from 2001 to 2010.

In 1998, Dr. Goswami was nominated for the Bliss Medal for Academic Excellence by the Society of American Military Engineers, and in 2003 he received the Educator of the Year Award from the Maryland Society of Professional Engineers. In 2016, he received the Excellence in Engineering Education award from the National Society of Professional Engineers.

CONTENTS

PREFACE

This book contains 280 questions divided into seven 40-question exams. There are two (Chapters 1 and 2) breadth (AM) exams representative of the NCEES guidelines for the Principles and Practice of Civil Engineering. The detailed solutions for these two exams are in Chapters 8 and 9, respectively. These exams are followed by five depth (PM) exams— Chapter 3 (Structural Depth), Chapter 4 (Geotechnical Depth), Chapter 5 (Water Resources & Environmental Depth), Chapter 6 (Transportation Depth), and Chapter 7 (Construction Depth). The detailed solutions for these exams are in Chapters 10–14, respectively.

There are two ways you can use this book. Early in your review process, you can use the entire set of 280 questions as a repository of practice problems. The questions have been formatted to be similar to those on the actual exam given by the NCEES. As with any other set of solved problems, remember that reading through *example solutions created by someone else is not adequate preparation for an exam such as this.*

I find that when I am reading the solution to an example, unless I police myself consciously, I sometimes don't think about *why the solution starts the way it does.* I find myself checking the arithmetic and that the numbers in step 1 lead to the numbers in step 2 and so on, but I don't ask myself, "Would I have started the solution the same way, based on the specifics of the question?" In any kind of problem solving, the initial setup of the solution is more than half the battle; the rest is usually mostly arithmetic and algebra.

A second way to utilize the book is to combine one of the breadth (AM) exams with the depth exam of your choice to create a full 8-hour exam. Take this timed exam toward the end of your review, when you feel you are adequately prepared. In the second edition, more than 80 of the 280 problems have been completely changed, some to focus on different types of problems, and others to reflect changes in the official NCEES syllabus.

The first edition of this book was focused largely on quantitative problems, whereas in this edition, a significant number of qualitative questions have been added. These could literally be from anywhere and how well you are able to answer them depends on (1) your familiarity with the applicable codes and standards and (2) your ability to "see the big picture" in a particular subject. This ability is what I like to call the *difference between knowledge and wisdom.* In order to develop this wisdom, you *must* supplement problem solving with readings from a few well-written and well-organized books. You probably need this level of wisdom only for your depth area, so allocate your limited time accordingly.

In using these and any other practice exams, you should develop and implement test-taking strategies such as making multiple passes through the exam without getting bogged down in unfamiliar areas, practicing effective time management, and effective use of tabbing all materials.

It is perhaps unavoidable that on an exam such as the *PE*, you will spend some time on a particular question before realizing that it is "not going well" and decide to move on to the next one. If you have prepared well, you should have a very good handle on your strengths and weaknesses. This will allow you to quickly recognize the intent of a question and make a quick decision on whether to attempt the problem right away, or whether to come back to it in the next pass. The quicker you make this decision, the better off you are in terms of effective use of time. Make sure you have some personal method to mark eliminated (or favored) choices before you move on, so that if you come back to this problem and are forced to make a guess, you make a higher probability guess.

I hope you find this book useful and I wish you good luck on the PE exam.

Indranil Goswami

ACKNOWLEDGMENTS

I would like to acknowledge several people without whose input this book could not have been written in its present form. As it stands, even after rigorous proofreading, I am sure there are instances of ambiguity or lack of elegance in the wording of some of the questions, but my primary purpose in writing this book has been to present questions in the formatting likely to be on the actual exam given by the NCEES. In content, I have tried to remain true to the official syllabus for the exam, as it stands today. I would sincerely appreciate reader feedback on these questions, so that they may be improved in later editions.

There are many who I should thank for their valuable input in the development of these questions. First and foremost, many thanks go out to Ravindra Koirala, who provided a lot of suggestions, particularly about the structural problems. Valuable suggestions have also come from those who have attended my review course. Without this input, many of the problems would have inconsistencies and ambiguities that would make them less valuable to the reader.

I am eternally grateful to the late Larry Hager, my first editor at McGraw Hill Professional, under whose watchful eye the first edition of this book was published. Even before I realized it, Larry saw the potential for the books and always remained accessible whenever I needed answers.

Thanks are also due to Warishree Pant and her team at KnowledgeWorks Global Limited for managing the editing, proofreading, and design of the final document.

1

Breadth Exam No. 1

The following set of questions numbered 1 to 40 is representative of a 4-hr breadth exam according to the syllabus and guidelines for the Principles and Practice (P&P) of Civil Engineering Examination administered by the National Council of Examiners for Engineering and Surveying (NCEES), current for the October 2020 examination.

Enter your answers on page 26. Detailed solutions are on page 177.

1

A 12-ft-high × 60-ft-long × 12-in-thick free-standing wall is to be constructed using wall forms that are 12 ft high × 20 ft long.

Labor cost for erecting forms = $4.30/ft² (new) and $1.30/ft² (re-use)
Labor cost for dismantling forms = $1.05/ft²
Cost of concrete (assume 10% waste) = $120/yd³
Cost of reinforcement = $25/yd³

The cost (labor + materials) of building the wall is most nearly:

A. $7,400
B. $8,100
C. $8,600
D. $9,000

2

A room is 35 ft × 25 ft in plan. Ceiling height is 14 ft. Openings for doors and windows total 85 ft². The following data is given for plastering and painting operations.

Plaster and paint crew:

1 supervisor $30/hr
1 laborer $12/hr
2 painters $18/hr

Plastering productivity = 50 ft²/L.H.

Painting productivity = 150 ft²/L.H.

The estimated labor cost for plastering and painting the room (walls and ceiling) is most nearly:

A. $1,020
B. $1,280
C. $1,460
D. $1,690

3

An activity on arrow network for a project is shown here. Numbers adjacent to arrows are activity durations (weeks). Assume project start date is week 0. Based on end-of-week calculations for starts and finishes, the early start date (week) for activity G is:

A. 9

B. 10

C. 11

D. 12

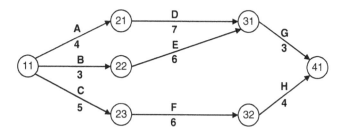

4

A project consists of 10 activities as outlined in the following table. All relationships are finish-to-start unless otherwise noted.

Activity	Duration (months)	Predecessors	Successors
A	5	—	C, D, E
B	3	—	H
C	3	A	F
D	4	A	G
E	3	A	H
F	5	C	J
G	4	D (FF LAG = 5)	J
H	3	B, E	I
I	2	H	J
J	2	F, G, I	—

(*Continued on next page*)

The activity on arrow representation of the project is also shown as follows.

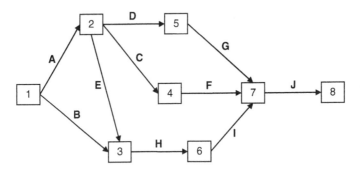

The critical path for the project is:

A. ACFJ

B. ADGJ

C. BHIJ

D. AEHIJ

5

During exercising project controls at a certain time for a project, the project manager computes the following:

BCWS = $435,000
ACWP = $510,000
BCWP = $488,000

Which of the following is true?

A. The project is ahead of schedule but overbudget.

B. The project is behind schedule and overbudget.

C. The project is ahead of schedule and underbudget.

D. The project is behind schedule but underbudget.

6

A contractor needs to bring in 4,200 yd³ of select soil to replace unsuitable subgrade material for a new highway. The borrow site is located 2.6 miles away with an average round trip travel/loading/dumping time of 30 min. The soil has a unit weight of 165 lb/ft³, and the dump truck drivers are on 10-hr workdays. The minimum number of 10-ton (net capacity) trucks needed to complete the job within 8 working days is:

A. 5
B. 6
C. 7
D. 8

7

A temporary warning sign is constructed at a worksite by using a 14-in-diameter bucket filled with ballast so as to serve as a counterweight (see figure). If the maximum (3-sec gust) wind pressure is 55 psf (ignore the wind pressure on the bucket), the minimum required weight (lb) of the counterweight is most nearly:

A. 165
B. 280
C. 330
D. 560

8

A cantilever-reinforced concrete retaining wall is shown. The friction angle between the wall footing and the soil is 20°. The factor of safety for sliding of the wall is most nearly:

A. 1.27
B. 0.93
C. 1.45
D. 1.78

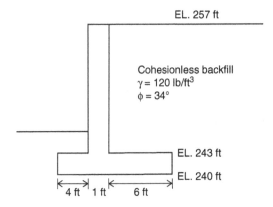

EL. 257 ft

Cohesionless backfill
$\gamma = 120$ lb/ft^3
$\phi = 34°$

EL. 243 ft

EL. 240 ft

4 ft   1 ft   6 ft

9

A plate load test is conducted on a sandy soil stratum. The plate is 12 in × 12 in. A concentric static load of 6,000 lb is applied to the plate. The soil has the following properties:

Unit weight = 125 lb/ft^3
Moisture content = 15%
Void ratio = 0.50
Modulus of elasticity = 300 kip/ft^2
Coefficient of subgrade reaction = 500 lb/in^3

The vertical settlement (inches) due to the application of the load is most nearly:

A. 0.02
B. 0.05
C. 0.08
D. 0.11

10

A direct shear test is performed on a soil sample. The sample is capped by porous end-plates and subjected to a normal stress in a split-box apparatus. The sample is then subjected to shear forces that induce shear failure on the interface plane. Results are summarized as follows:

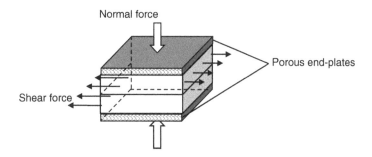

Sample A tested at a total vertical stress equal to 1,000 psf failed at a shear stress of 675 psf; while sample B tested at a total vertical stress of 3,000 psf failed at a shear stress of 2,025 psf.

Based upon the results of this direct shear test, the shear strength parameters for this soil are most nearly:

A. $c = 0$ psf and $\phi = 34°$
B. $c = 200$ psf and $\phi = 34°$
C. $c = 0$ psf and $\phi = 42.5°$
D. $c = 200$ psf and $\phi = 42.5°$

11

A clay soil deposit has the following characteristics:

Unit weight = 122 lb/ft^3
Angle of internal friction = 10°
Unconfined compression strength = 2,400 lb/ft^2

A square footing is to be used to support a column load = 80 k. Bottom of footing is 3 ft below the ground surface. Minimum factor of safety for ultimate bearing capacity is 3.0. The minimum footing size required (ft) is most nearly:

A. 6
B. 5
C. 4
D. 3

12

A boring log for a sewer rehabilitation project is shown here. Assume the following total unit weights:

All clays and silts: 130 lb/ft^3
All sands: 120 lb/ft^3

The standard penetration resistance, after correcting for overburden pressure, for split-spoon sample no. 5 is most nearly:

A. 20
B. 25
C. 30
D. 35

RECORD OF SOIL/ROCK EXPLORATION

Contracted with _____ Boring # ___SB-1___

Project Name ___Power Mil Sewer and Stream Project_____ Job # ___16-00053.02a___

Location ___Baltimore City, Maryland_____

SAMPLER

Datum _____   Hammer Wt ___140 lb___   Hole Diameter ___6.25 in___   Foreman ___B. Jackson___

Surf. Elev. ___368.0 ± ft___   Hammer Drop ___30 in___   Rock Core Dia. ___N/A___   Inspector _____

Date Started ___8/15/16___   Spoon Size ___2 in___   Boring Method ___HSA___   Data Completed ___8/15/16___

ELEV. (ft)	SOIL DESCRIPTION Color, Moisture, Density, Plasticity, Size Proportions	STRATUM DEPTH (ft)	SOIL SYMBOL	DEPTH SCALE	SAMPLE					BORING & SAMPLE NOTES
					Cond	Blows/6"	No.	Type	Rec (in)	
367.6	5-inches of TOPSOIL	0.4								1. Boring offset at 8.0-ft West due to existing sewer line.
	Brown, damp, stiff, Silty CLAY, (cl)				I	3-4-5	1	DS	14	
365.0		3.0								2. Bulk Sample taken 6 to 10-ft
	Red-brown, damp, very stiff, Silty CLAY, trace Sand (d)			5	I	6-11-14	2	DS	16	
362.0		6.0								
	Gray-brown, damp to most, medium dense, Silty SAND, And Sandy SILT (sm)				I	6-8-9	3	DS	18	
				10	I	7-8-9	4	DS	12	
355.0		13.0								
	Dark gray, brown, moist, medium dense, Silty SAND (sm)			15	I	6-11-14	5	DS	18	
348.0		20.0		20	I	6-7-9	6	DS	15	
	Bottom of Boring at 20.0 ft			25						

SAMPLER TYPE	SAMPLE CONDITIONS	GROUNDWATER DEPTH	BORING METHOD
DS - DRIVEN SPLIT SPOON	D - DISINTEGRATED	AT COMPLETION _130_ ft	HSA - HOLLOW STEM AUGERS
PT - PRESSED SHELBY TUBE	I - INTACT	AFTER ___ HRS ___ ft	CFA - CONTINUOUS FLIGHT AUGERS
CA - CONTINUOUS FLIGHT AUGER	U - UNDISTURBED	AFTER 24 HRS _8.0_ ft	DC - DRIVING CASING
RC - ROCK CORE	L - LOST	CAVED AT _13.0_ ft	MD - MUD DRILLING

STANDARD PENETRATION TEST DRIVING 2" OD SAMPLER 1' WITH 140W HAMMER FALLING 30". COUNT MADE AT 6" INTERVALS

13

A 6-in-thick riprap layer is used as protection for the earth slope ($\theta = 30°$) as shown here. The factor of safety for slope stability is most nearly:

A. 1.25
B. 1.35
C. 1.45
D. 1.55

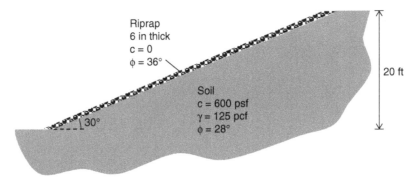

Riprap
6 in thick
c = 0
$\phi = 36°$

20 ft

Soil
c = 600 psf
$\gamma = 125$ pcf
$\phi = 28°$

30°

14

For the beam loaded as shown here, the distance (ft), measured from the left support, at which the transverse shear force = 0 is most nearly:

A. 5.5
B. 8.5
C. 11.5
D. 14.5

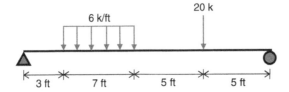

20 k

6 k/ft

3 ft   7 ft   5 ft   5 ft

15

For the plane truss shown here, the force (lb) in member BE is most nearly:

A. 27,000 (compression)
B. 27,000 (tension)
C. 41,000 (compression)
D. 41,000 (tension)

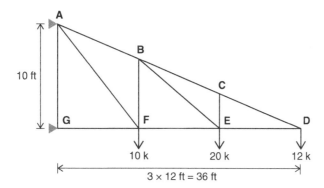

16

A rectangular cantilever beam is bent into an L-shape as shown here. A force P = 10 k acts on the beam as shown. The maximum normal stress (lb/in²) is most nearly:

A. 8,600
B. 9,000
C. 9,500
D. 9,800

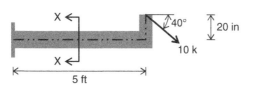

Section X-X

17

A hoisting mechanism uses a cable system as shown here. If the load $W = 3$ tons, the effort required (lb) at point B is most nearly:

A. 1,345
B. 1,715
C. 2,065
D. 2,350

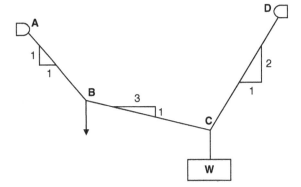

18

Steel bridge girders are set in place on their bearings when the ambient temperature is 65°F. If the temperature of the steel is expected to reach the extremes of 15°F and 95°F, what is most nearly the expansion gap (inches) that must be provided at the bearings?

The coefficient of thermal expansion for steel: $\alpha_s = 7.3 \times 10^{-6}/°F$

A. 1/8
B. 1/4
C. 3/4
D. 15/16

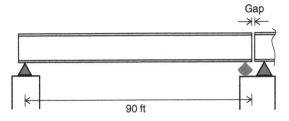

19

A simply supported steel beam (span = 25 ft) supports a uniformly distributed load = 2.75 k/ft. The allowable bending stress for the chosen grade of steel is 32 ksi. The lightest W-section (choose from the following table) is:

A. W12 × 50
B. W12 × 53
C. W12 × 58
D. W12 × 65

Section	A_g (in²)	d (in)	S_x (in³)	S_y (in³)
W12 × 79	23.2	12.4	107.0	35.8
W12 × 72	21.1	12.3	97.4	32.4
W12 × 65	19.1	12.1	87.9	29.1
W12 × 58	17.0	12.2	78.0	21.4
W12 × 53	15.6	12.1	70.6	19.2
W12 × 50	14.6	12.2	64.2	13.9

20

Water flows in a rectangular open channel at a normal depth of 4.6 ft as shown. Assume Manning's roughness coefficient (constant with depth) = 0.014. The longitudinal slope of the channel bed is 0.4%. The flow rate (ft³/sec) is most nearly:

A. 600
B. 700
C. 800
D. 900

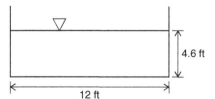

21

A detention pond receives flow from a mixed-use subdivision. Flow is allowed to discharge from the pond through a rectangular weir (width = 5 ft, crest elevation = 124.72 ft above sea level). When the surface elevation of the pond is 127.50 ft, the instantaneous rate of discharge through the weir (MGD) is most nearly:

A. 50
B. 63
C. 77
D. 92

22

A watershed (area = 370 acres) is subdivided into five distinct land use classifications, as shown in the following table. Storm runoff data have been abstracted into a set of intensity–duration–frequency curves. Using the Rational Method, the runoff discharge (ft³/sec) from a 20-year storm with gross rainfall = 5.6 in is most nearly:

A. 300
B. 600
C. 900
D. 1200

Region	Area (acres)	Land Use	Soil Type	Time for Overland Flow (min)	Curve Number	Rational Runoff Coefficient
A	80	Lawns: fair condition	B	30	69	0.4
B	80	Forest	C	45	45	0.2
C	50	Paved	B	15	98	0.9
D	90	Residential: 4 lots per acre	D	25	87	0.6
E	70	Forest	A	45	35	0.2

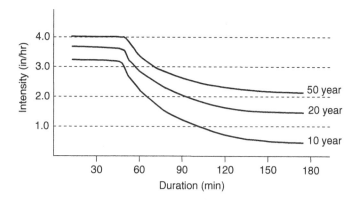

23

The 1-hr unit hydrograph of excess precipitation is described by the following data:

Time (hr)	0.0	1.0	2.0	3.0	4.0	5.0
Discharge Q (ft³/s/in)	0	30	95	125	50	0

A storm produces the following pattern of excess precipitation—1.7 in of excess precipitation during the first hour, followed by 0.8 in during second hour. The stream discharge (ft³/s) at the end of the second hour is most nearly:

A. 140
B. 155
C. 170
D. 185

24

A catchment area is divided into eight Thiessen polygons. Each polygon corresponds to a rain gage station, as summarized in the following table. The average precipitation (inches) for the catchment area is most nearly:

A. 1.3
B. 1.4
C. 1.5
D. 1.8

Thiessen Polygon #	Rain Gage Station ID	Area (A)	Depth D (in)
012891	A-1	2.1	1.7
024119	A-2	3.2	2.1
039003	A-3	5.2	1.9
412002	B-1	2.9	1.2
444511	B-2	4.9	1.1
451903	B-3	6.1	0.9
478122	B-4	3.1	1.5
490021	B-5	2.2	2.0

25

A 2:1 reducer in a horizontal pipe with an upstream diameter of 6 in conveys flow of 950 gal/min. If energy losses in the reducer can be ignored, the pressure loss (psi) in the reducer is most nearly:

A. 34
B. 27
C. 17
D. 12

26

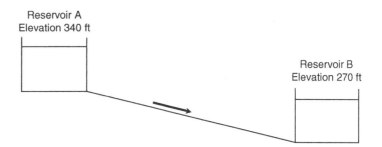

Reservoir A
Elevation 340 ft

Reservoir B
Elevation 270 ft

Water flows by gravity from reservoir A (surface elevation 340 ft) to reservoir B (surface elevation 270 ft) through a system of cast-iron pipes. The characteristics of the pipe system are listed here:

Length = 2,500 ft
Diameter = 24 in
Friction factor = 0.02
Minor loss equivalent length = 55 ft

The flow rate (ft³/sec) is most nearly:

A. 42
B. 55
C. 65
D. 76

27

The coordinates (ft) of the PC and PI for a circular horizontal curve are given here:

PC: 1232.56 N, 123.32 E
PI: 509.72 N, 172.11 W

The degree of curve = 4 degrees

The curve deflects to the left.

The coordinates of the PT (ft) are:

A. 256.21 N, 121.72 W
B. 1157.28 S, 811.83 W
C. 1149.62 N, 619.67 E
D. 130.18 S, 275.45 E

28

A parabolic vertical curve is to connect a tangent of +5% to a gradient of −4%. If the PVI is a station 123 + 32.50 and the tangent offset at the PVT is 17.65 ft, the station of the PVC is most nearly:

A.   117 + 12.25
B.   119 + 40.28
C.   119 + 24.21
D.   121 + 36.39

29

A stadium hosts an event with an audience of 40,000. Approximately 30% of the audience are expected to use an adjacent light rail station following the event. It is anticipated that about 90% of the stadium empties in the first hour after the conclusion of the event. A dedicated pedestrian walkway connects the stadium to the light rail station. The effective width of the walkway is 32 ft. Assume the PHF (based on peak 15-min flow) for the walkway is 0.88. The peak flow rate (ped/min/ft) on the walkway during the first hour is most nearly:

A.   4.8
B.   5.6
C.   6.4
D.   7.2

30

Grain size distribution for a soil sample is shown in the following curve. Liquid limit = 34%. Plastic limit = 19%. What is the USCS classification?

A. SW
B. SP
C. ML
D. CL

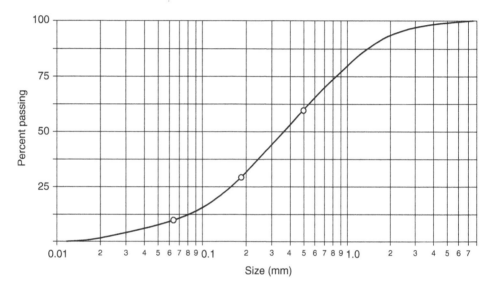

31

A soil sample weighs 3.64 lb and has volume = 0.031 ft³. Specific gravity of soil solids = 2.65. Water is added to the soil until it bleeds. The added water has volume = 5.6 fl.oz. The dry unit weight of the original soil sample (lb/ft³) is most nearly:

A. 101
B. 104
C. 107
D. 110

32

A concrete mix is proportioned 1 : 1.8 : 2.6 (cement : sand : coarse aggregate) by weight. The following specifications are given:

Cement	specific gravity = 3.15
SSD sand	(m.c. = 0.5%) specific gravity = 2.70
SSD coarse aggregate	(m.c. = 0.7%) specific gravity = 2.60
Added water	5.8 gal per sack cement
Air	3% (by volume)

The aggregates used for mixing the concrete had the following properties:

Wet sand:	moisture content = 6%
Wet coarse aggregate:	moisture content = 4%

The water content (gal/sack) of the concrete is most nearly:

A. 6

B. 7

C. 8

D. 9

33

The yield strain (%) of A36 steel is most nearly:

A. 0.250

B. 0.125

C. 0.095

D. 0.055

34

A tensile test coupon, cut from ¼-in-thick steel plate, is being tested in a Universal Testing Machine. The dimensions of the specimen are shown here. The stress–strain diagram of the steel is also shown.

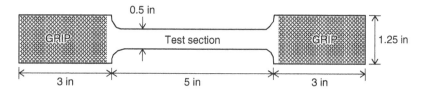

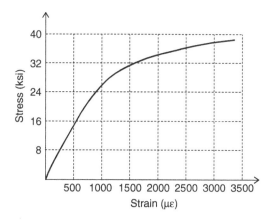

The axial force (lb) at which the yielding of the specimen begins is most nearly:

A. 2,400
B. 4,700
C. 6,100
D. 7,200

35

A soil sample, taken from a borrow pit has a specific gravity of soil solids of 2.66. Six samples with varying moisture content were prepared for the Standard Proctor test. The results of the standard Proctor test are shown in the following table.

Sample	Weight of Soil (lb)	Water Content (%)
1	3.20	12.8
2	3.78	13.9
3	4.40	15.0
4	4.10	15.7
5	3.70	16.6
6	3.30	18.1

The maximum dry unit weight (lb/ft^3) is most nearly:

A.   85

B.   90

C.   100

D.   115

36

The following table shows cross-section areas of cut and fill recorded at five stations spaced at 100 ft. Shrinkage is 15%.

Station	Area (ft^2) Cut	Fill
0 + 0.00	245.0	423.5
1 + 0.00	312.5	176.3
2 + 0.00	111.5	303.0
3 + 0.00	234.5	188.4
4 + 0.00	546.2	514.5

If the ordinate of the mass diagram at station $0 + 00$ is $+400$ yd³, the ordinate (yd³) at station $4 + 00$ is most nearly:

A. −650
B. −1,450
C. −27,920
D. +28,720

37

A construction stake has the markings shown here. Which of the following is NOT a correct interpretation of this stake?

A. The elevation of the flow line invert is 823.23 ft.
B. This is a drainage stake for plan feature 13-A.
C. The centerline of the grate is 15.35 ft from the stake.
D. The stake is along the flow line of a dike.

DS 13-A
FL DIKE
EL. 823.23
15^{35} CL GRT

38

Tracking of sediment from a construction site can be controlled by using a stabilized construction entrance. The recommended configuration is:

A. Minimum length 30 ft, 2 to 3 in stone over sod
B. Minimum length 30 ft, ¾ in nominal size stone over geotextile fabric
C. Minimum length 50 ft, ¾ to 1 in nominal size stone over geotextile fabric
D. Minimum length 50 ft, 2 to 3 in nominal size stone over geotextile fabric

39

Which of the following techniques are commonly used for construction adjacent to historic structures?

    I. Underpinning
   II. Anchor rod and deadman
  III. Slurry walls
  IV. Compaction piles

A. I and II
B. I and III
C. II and III
D. II and IV

40

A structure has been examined to determine a need for rehabilitation. The related costs are summarized as follows:

Current annual costs = $40,000
Estimated rehabilitation cost = $350,000
Annual costs projected after rehabilitation = $15,000
Expected useful life remaining = 20 years
Projected increase in residual value (at end of useful life) = $200,000

The return on investment (ROI) for performing the rehabilitation is most nearly:

A. 4%
B. 5%
C. 6%
D. 7%

END OF BREADTH EXAM NO. I

USE THE ANSWER SHEET ON THE NEXT PAGE

Breadth Exam No. 1: Answer Sheet

1	(A)	(B)	(C)	(D)	21	(A)	(B)	(C)	(D)
2	(A)	(B)	(C)	(D)	22	(A)	(B)	(C)	(D)
3	(A)	(B)	(C)	(D)	23	(A)	(B)	(C)	(D)
4	(A)	(B)	(C)	(D)	24	(A)	(B)	(C)	(D)
5	(A)	(B)	(C)	(D)	25	(A)	(B)	(C)	(D)
6	(A)	(B)	(C)	(D)	26	(A)	(B)	(C)	(D)
7	(A)	(B)	(C)	(D)	27	(A)	(B)	(C)	(D)
8	(A)	(B)	(C)	(D)	28	(A)	(B)	(C)	(D)
9	(A)	(B)	(C)	(D)	29	(A)	(B)	(C)	(D)
10	(A)	(B)	(C)	(D)	30	(A)	(B)	(C)	(D)
11	(A)	(B)	(C)	(D)	31	(A)	(B)	(C)	(D)
12	(A)	(B)	(C)	(D)	32	(A)	(B)	(C)	(D)
13	(A)	(B)	(C)	(D)	33	(A)	(B)	(C)	(D)
14	(A)	(B)	(C)	(D)	34	(A)	(B)	(C)	(D)
15	(A)	(B)	(C)	(D)	35	(A)	(B)	(C)	(D)
16	(A)	(B)	(C)	(D)	36	(A)	(B)	(C)	(D)
17	(A)	(B)	(C)	(D)	37	(A)	(B)	(C)	(D)
18	(A)	(B)	(C)	(D)	38	(A)	(B)	(C)	(D)
19	(A)	(B)	(C)	(D)	39	(A)	(B)	(C)	(D)
20	(A)	(B)	(C)	(D)	40	(A)	(B)	(C)	(D)

2

Breadth Exam No. 2

The following set of questions numbered 101 to 140 is representative of a 4-hr breadth exam according to the syllabus and guidelines for the Principles and Practice (P&P) of Civil Engineering Examination administered by the National Council of Examiners for Engineering and Surveying (NCEES), current for the October 2020 examination.

Enter your answers on page 52. Detailed solutions are on page 195.

101

A grade beam 3 ft wide by 2 ft deep must be poured around the perimeter of a building whose plan is shown in the following figure. The trench excavated for the grade beam has to have 1V:1H side slopes. Excavation is accomplished by a trackhoe, having productivity of 9 yd³/hr. Assume an 8-hr workday.

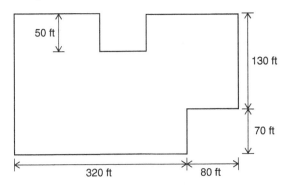

The number of days to complete the excavation activity is most nearly:

A. 4
B. 5
C. 6
D. 7

102

The tasks within a project and their duration and cost data are shown below. Both normal completion time and crashed completion time (by allocating additional resources) are shown in the following table.

Activity	Normal Time (weeks)	Normal Cost	Crash Time (weeks)	Crash Cost
A	3	$3,000	2	$5,000
B	4	$4,000	2	$6,000
C	5	$5,000	3	$8,000
D	8"	$5,000	6	$6,000
E	3"	$3,000	2	$4,000
F	5"	$4,000	3	$8,000

The activity on arrow representation of the project is also shown in the following figure.

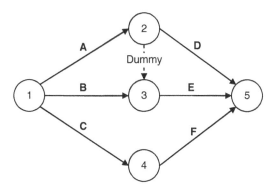

For every week of early completion, the bonus is $1,000, and for every week that the project is late, the penalty is $800. The net added cost to complete the project by the revised completion time of 9 weeks is most nearly:

A. $22,000
B. $24,000
C. $25,000
D. $26,000

103

The activity-on-node diagram for a project consisting of seven activities A through G is shown in the following figure. Numbers adjacent to each activity label are activity durations in days. All relationships are finish-to-start, except for a FF LAG = 5 between activities C and D. The minimum time to complete the project is most nearly:

A. 21
B. 20
C. 19
D. 18

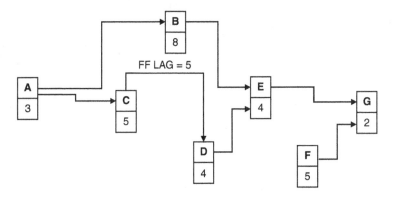

104

Which of the following chemicals is commonly used for dust control at construction sites?

A. Calcium chloride
B. Sodium thiosulfate
C. Calcium bicarbonate
D. Sodium phosphate

105

A crane is used to lift a load of 16 tons as shown. The crane cabin has a ballasted weight of 24 tons. The allowable soil pressure is 2,800 lb/ft². If the crane is supported by four outriggers as shown, the minimum contact area (ft²) of the outrigger pads is most nearly:

A. 9.0

B. 12.3

C. 16.0

D. 20.3

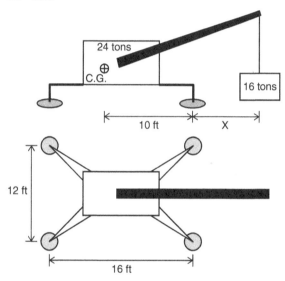

106

A landfill accepts municipal solid waste from town A (population 20,000). The landfill has been in operation for 6 years and has the remaining capacity of 1 million yd³. A new subdivision (population 5,000) is developed and populated. The waste generation rate is 5 lb/capita-day. The average compacted density in the landfill is 40 lb/ft³. The reduction in the service life of the landfill due to the addition of the new subdivision is most nearly:

A. 12 years

B. 10 years

C. 8 years

D. 6 years

107

A precast wall panel 20 ft long (perpendicular to drawing plane) × 14 ft high × 4 in thick is being tilted into position using a single horizontal cable connected to the top of the panel as shown in the following figure. The unit weight of the wall material is 80 lb/ft³. At the position shown, the tension in the cable (k) is most nearly:

A. 2.2
B. 3.6
C. 5.2
D. 6.6

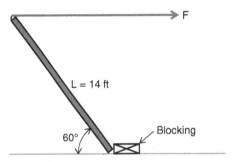

108

A cantilever-retaining wall is shown. The active resultant per unit length (lb/ft) of the wall is most nearly:

A. 280
B. 1,900
C. 4,260
D. 8,530

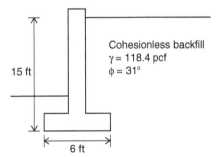

109

A clay deposit is 32 ft thick. It has the following properties: $\gamma = 126$ pcf, moisture content = 12%, void ratio = 0.60, $C_c = 0.20$, $C_r = 0.05$. If building on the surface causes the effective stress on the clay layer to double, the ultimate settlement (inches) due to consolidation is most nearly:

A. 14.4
B. 10.0
C. 5.1
D. 3.5

110

A 10-ft-high masonry wall erected at a construction site must be braced overnight for stability against an equivalent static wind pressure of 28 psf. The bottom of the wall is supported continuously by blocking, and timber braces are used to provide temporary bracing to the wall. If the axial compression capacity of each brace is 2,350 lb, then the maximum allowable spacing S (ft) is most nearly:

A. 13
B. 10
C. 8
D. 5

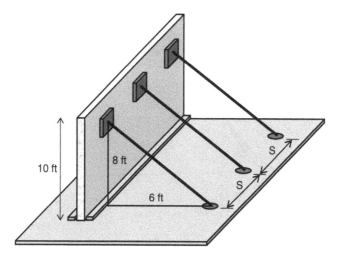

III

A 5-ft × 5-ft square footing transfers a column load of 140 k to a sandy soil as shown. The depth of the footing is 3 ft. The factor of safety against general bearing capacity failure is most nearly:

A. 1.3
B. 2.0
C. 2.7
D. 3.3

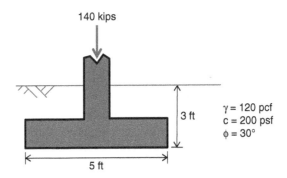

140 kips

3 ft

γ = 120 pcf
c = 200 psf
ϕ = 30°

5 ft

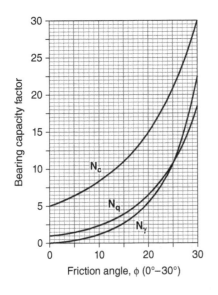

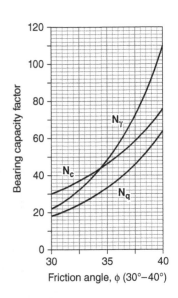

112

In a triaxial test, a cylindrical (2 in diameter, 4 in length) clay sample is tested to failure. Before the axial load is applied, the sample is subjected to a hydrostatic chamber pressure of 15 lb/in². The sample fails in shear when the added axial load is 38 lb. At this time, the pore pressure is measured to be 6.5 lb/in². The cohesion (lb/ft²) of the soil is most nearly:

A. 1,000

B. 2,100

C. 3,000

D. 4,200

113

A reinforced-concrete cantilever wall is used to retain an engineered fill as shown. Drains are designed so that water pressure does not build up behind the wall. Which of the following patterns shows the correct placement of flexural reinforcement in the wall?

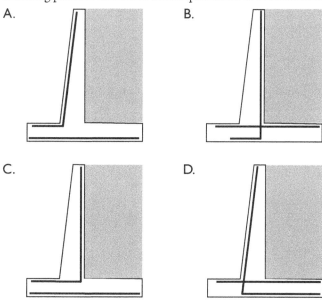

114

For the truss shown in the following figure, the member force (k) in member DI is most nearly:

A. 22 (compression)
B. 22 (tension)
C. 39 (compression)
D. 39 (tension)

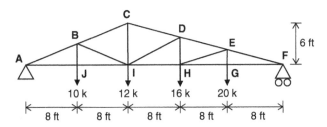

115

A steel beam has a singly symmetric I-shaped cross section as shown. The elastic section modulus (in³) about the major centroidal axis is most nearly:

A. 54
B. 63
C. 66
D. 73

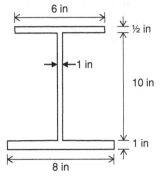

116

A residential structure has a floor supported by a system of joists spaced every 36 in as shown. The combined floor load is 90 psf. The allowable bending stress in the timber joists is 1,700 psi. The required section modulus (in³) of the timber joists is most nearly:

A. 30

B. 50

C. 80

D. 120

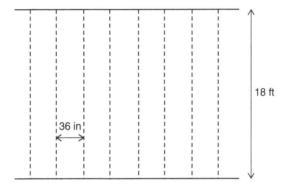

117

A noncomposite bridge girder (65-ft simple span) carries a uniformly distributed load = 4.75 k/ft. If the maximum allowable deflection is L/360, the required moment of inertia (in⁴) is most nearly:

A. 41,200

B. 30,500

C. 25,800

D. 20,200

118

For the cable system shown, the tension (k) in cable AB is most nearly:

A. 16
B. 19
C. 22
D. 25

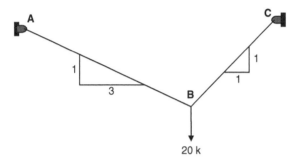

119

Timber studs (Douglas Fir-Larch) spaced every 30 in form a load-bearing wall. The vertical load is transferred to the studs by a header beam, as shown in the following figure. The studs have a rectangular cross section 1.75 in × 3.25 in, with the 1.75 in edge parallel to the length dimension of the wall. Nails are used to connect the sheathing to each face of the stud. If the height of the wall is 10 ft 6 in, the Euler buckling load (lb) of each stud is most nearly:

A. 3,230
B. 3,505
C. 4,105
D. 4,710

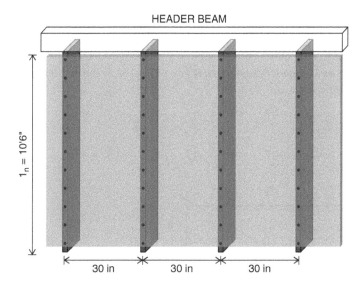

HEADER BEAM

$1_n = 10'6''$

30 in   30 in   30 in

120

A flow rate = 20 ft³/sec flows through a 30-in diameter concrete pipe (Manning's n constant with depth = 0.013). The following data is given:

Pipe length = 800 ft
Pipe invert elevation at upstream end of pipe = 275.64 ft
Pipe invert elevation at downstream end of pipe = 270.96 ft

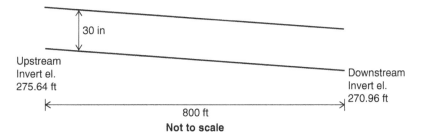

30 in

Upstream
Invert el.
275.64 ft

Downstream
Invert el.
270.96 ft

800 ft
Not to scale

The depth of flow (inches) in the pipe is most nearly:

A. 14.0
B. 15.3
C. 17.5
D. 21.8

121

A trapezoidal channel has bottom width = 2 ft, a longitudinal slope of 0.5% and sides at 1V:3H slopes as shown. Manning's n is given as 0.020. If the depth of flowing water is 2 ft, the velocity (fps) is most nearly:

A. 5.6
B. 6.8
C. 3.5
D. 3.8

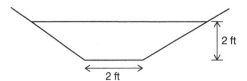

122

Normal depth of flow occurs in a rectangular open channel. The following data is given:

Flow rate = 1,200 cfs
Manning's roughness coefficient = 0.015
Bottom width = 10 ft
Longitudinal slope = 1%

The Froude number is most nearly:

A. 0.5
B. 1.4
C. 3.9
D. 7.2

123

The 1-hr unit hydrograph of excess precipitation is shown in the following table.

Time (hr)	0	1	2	3	4	5
Discharge Q (cfs/in)	0	35	75	105	40	0

A 2-hr storm produces 1.7 in of runoff during the first hour followed by 0.8 in of runoff during second hour. The peak discharge (ft^3/sec) due to this storm is most nearly:

A. 210
B. 239
C. 263
D. 287

124

Which of the following statements is/are FALSE?

    I. Flood stage (elevation) increases with an increase in the return period.
    II. Annual probability of occurrence of a 20-year storm is half of that for a 10-year storm.
    III. The 50-year flood is guaranteed to occur at least once in a 100-year interval.
    IV. Return period of design event does not directly impact the design.

A. I and III
B. II and III
C. III only
D. III and IV

125

A watershed is subdivided into five distinct land-use/land-cover as summarized in the following table. The intensity–duration–frequency curves are synthesized from regional storm records.

Region	Land use/ Land Cover	Area (acres)	Rational Coefficient	Time of concentration (min)
A	Wooded and forested	50	0.20	40
B	Residential lots	65	0.55	30
C	Parking lots	20	0.85	25
D	Lawns	240	0.30	35

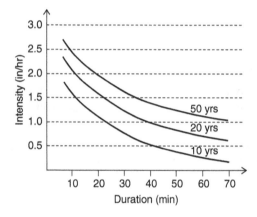

The discharge (ft³/sec) due to a 50-year storm is most nearly:

A. 70

B. 110

C. 190

D. 260

126

A 12-in pipe (cast iron, Hazen Williams $C = 120$) serves as the outfall from a large reservoir whose water surface elevation is 145 ft. The average longitudinal slope of the pipe is 0.01 ft/ft. The bottom of the reservoir is at elevation 125 ft, and the outfall of the pipe is at elevation 95 ft. Kinematic viscosity of water is 1.217×10^{-5} ft²/sec. The flow rate (gal/min) in the pipe is most nearly:

A. 1,000
B. 1,500
C. 2,000
D. 2,500

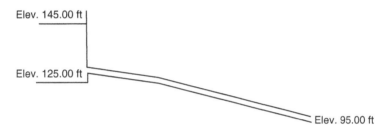

127

A horizontal circular curve has PC at coordinates (ft) 4123.64 N, 1064.32 W. Curve radius = 1030 ft. The tangent at the PC has bearing S42°30′ W. The length of curve = 646.35 ft. The coordinates of the PI are:

A. 3877.23 N, 1290.12 W
B. 3897.84 N, 1310.73 W
C. 4370.05 N, 838.52 W
D. 4349.44 N, 817.91 W

128

A parabolic vertical curve joins a grade of −4% to a grade of +6%. The PVI is at station 10 + 56.30. Elevation of the PVI is 432.65 ft. The curve passes under a bridge structure at station 12 + 00.00. The elevation of the bottom of the bridge girders is 470.00 ft. The minimum vertical clearance under the bridge is 14 ft 6 in. The required length of curve (ft) is most nearly:

A. 1,665

B. 1,865

C. 2,065

D. 2,265

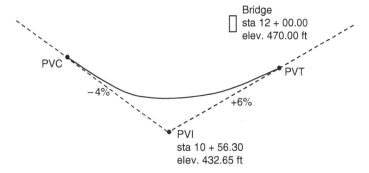

129

The following data is given for a six-lane freeway:

AADT = 78,500 vpd

K = 0.11

D = 0.57

PHF = 0.94

The peak design flow rate (veh/hr/ln) is most nearly:

A. 1,650

B. 1,750

C. 1,850

D. 1,950

130

Grain size distribution for a soil sample is shown in the following curve. Liquid limit = 34%. Plastic limit = 19%. What is the AASHTO classification?

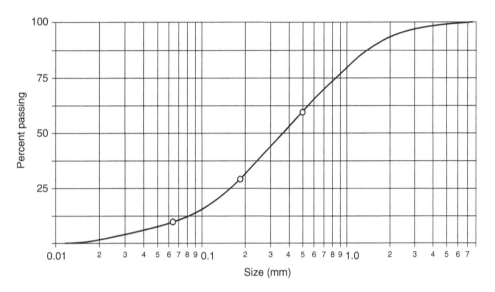

A. A-6

B. A-2-6

C. A-2-7

D. A-7

131

Results from sieve analysis of a soil sample have been summarized in the following table. Atterberg tests resulted in: liquid limit = 54, plastic limit = 23. What is the USCS classification for the soil?

A. GP
B. GP-GC
C. GP-GM
D. GW-GC

Sieve Size	% Passing
2 in	95
1 in	85
½ in	60
No. 4	41
No. 10	31
No. 40	22
No. 200	10

132

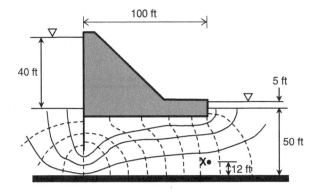

A gravity dam has a cutoff wall at the upstream end as shown in the figure. The width of the base of the dam is 100 ft and the length of the dam is 150 ft. An impermeable layer exists approximately 50 ft below the bottom of the reservoir. The upstream reservoir depth is 40 ft and the tailwater depth is 5 ft. The hydraulic conductivity of the soil in the

50-ft.-thick soil layer underlying the dam is 0.0025 ft/sec. The pressure (psig) at point X (located 12 ft above the top of the impermeable layer) is most nearly:

A. 45

B. 37

C. 30

D. 22

133

Which of the following statements is correct?

I.   Air entrainment always lowers concrete strength.

II. Air entrainment increases concrete durability.

III. Air entrainment is always desirable in concrete structures.

IV. The amount of entrained air depends on maximum aggregate size.

A. II and III

B. II and IV

C. I and III

D. I, II, and IV

134

The design target 28-day compressive strength of concrete for a bridge construction project is 6,000 psi. Concrete samples will be tested as part of the QA/QC efforts that are part of the project. The sample standard deviation is 675 psi.

The minimum required average compressive strength (psi) to be used in concrete mix design is most nearly:

A. 6,500

B. 6,750

C. 7,000

D. 7,250

135

Which of the following is/are NOT an example of soil testing?

I. Ground penetrating radar
II. Nuclear density test
III. Brinell hardness test
IV. Liquid penetrant test
V. Cone penetrometer test

A. III and IV
B. II, III, and V
C. IV and V
D. II and V

136

During repaving of a highway segment, project specifications dictate that a buffer zone of width 12 ft on either side of the 65-ft-wide roadway be subject to clearing and grubbing. Project limits (measured along baseline curve along centerline of roadway) are from station 5 + 05.25 to 25 + 31.20. Unit cost for clearing and grubbing is $0.20 per SF.

The total line item cost for "Clearing and Grubbing" is most nearly:

A. $10,000
B. $12,000
C. $14,000
D. $16,000

137

A lath stake with the following markings is observed adjacent to a pipe feature:

12^{00}
$21 + 50$
12×36
CMP w/2
FL END

Which interpretation is NOT consistent with these markings?

A. The reference point is 12 ft from the pipe.
B. The pipe is 12 ft long and 36 in diameter.
C. The pipe is reinforced concrete pipe.
D. The pipe has two flared ends.

138

An embankment is to be constructed by placing fill in a 300-ft-long section of a site. Areas of fill sections 50 ft apart are shown in the following table. The total volume of earthwork (yd^3) between stations $12 + 0.00$ and $15 + 0.00$ is most nearly:

A. 4,030
B. 8,060
C. 12,100
D. 16,100

Station	Fill Area (ft²)
12 + 00.00	456.33
12 + 50.00	563.97
13 + 00.00	702.24
13 + 50.00	1234.98
14 + 00.00	783.92
14 + 50.00	591.94
15 + 00.00	493.34

139

The following figure shows a bridge abutment located adjacent to a roadway. The underlying coordinate grid (ft) has its origin $(0, 0)$ at the interface of the pavement and the face of the abutment wall. The bottom of the footing is located at a depth of 6.5 ft below grade. A sanitary sewer must be installed via open cut. The bottom of the trench is at a depth of 12.0 ft. The dashed line shows the trench outline to be located at an undetermined distance (X) from the face of the abutment wall. Without taking special measures to support the abutment, the minimum distance X (ft) from the face of the abutment wall to the edge of the trench is most nearly:

A. 4
B. 5
C. 6
D. 7

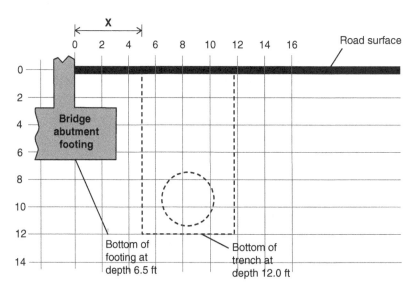

NOTE: The location of the edge of the trench is shown at 5 ft, for illustration purposes only.

140

A freestanding masonry wall of height 16 ft is being constructed. During construction, the width of the "restricted zone" should extend:

A. 18 ft on one side of the wall
B. 18 ft on both sides of the wall
C. 20 ft on one side of the wall
D. 20 ft on both sides of the wall

END OF BREADTH EXAM NO. 2

USE THE ANSWER SHEET ON THE NEXT PAGE

Breadth Exam No. 2: Answer Sheet

101	Ⓐ	Ⓑ	Ⓒ	Ⓓ
102	Ⓐ	Ⓑ	Ⓒ	Ⓓ
103	Ⓐ	Ⓑ	Ⓒ	Ⓓ
104	Ⓐ	Ⓑ	Ⓒ	Ⓓ
105	Ⓐ	Ⓑ	Ⓒ	Ⓓ
106	Ⓐ	Ⓑ	Ⓒ	Ⓓ
107	Ⓐ	Ⓑ	Ⓒ	Ⓓ
108	Ⓐ	Ⓑ	Ⓒ	Ⓓ
109	Ⓐ	Ⓑ	Ⓒ	Ⓓ
110	Ⓐ	Ⓑ	Ⓒ	Ⓓ
111	Ⓐ	Ⓑ	Ⓒ	Ⓓ
112	Ⓐ	Ⓑ	Ⓒ	Ⓓ
113	Ⓐ	Ⓑ	Ⓒ	Ⓓ
114	Ⓐ	Ⓑ	Ⓒ	Ⓓ
115	Ⓐ	Ⓑ	Ⓒ	Ⓓ
116	Ⓐ	Ⓑ	Ⓒ	Ⓓ
117	Ⓐ	Ⓑ	Ⓒ	Ⓓ
118	Ⓐ	Ⓑ	Ⓒ	Ⓓ
119	Ⓐ	Ⓑ	Ⓒ	Ⓓ
120	Ⓐ	Ⓑ	Ⓒ	Ⓓ

121	Ⓐ	Ⓑ	Ⓒ	Ⓓ
122	Ⓐ	Ⓑ	Ⓒ	Ⓓ
123	Ⓐ	Ⓑ	Ⓒ	Ⓓ
124	Ⓐ	Ⓑ	Ⓒ	Ⓓ
125	Ⓐ	Ⓑ	Ⓒ	Ⓓ
126	Ⓐ	Ⓑ	Ⓒ	Ⓓ
127	Ⓐ	Ⓑ	Ⓒ	Ⓓ
128	Ⓐ	Ⓑ	Ⓒ	Ⓓ
129	Ⓐ	Ⓑ	Ⓒ	Ⓓ
130	Ⓐ	Ⓑ	Ⓒ	Ⓓ
131	Ⓐ	Ⓑ	Ⓒ	Ⓓ
132	Ⓐ	Ⓑ	Ⓒ	Ⓓ
133	Ⓐ	Ⓑ	Ⓒ	Ⓓ
134	Ⓐ	Ⓑ	Ⓒ	Ⓓ
135	Ⓐ	Ⓑ	Ⓒ	Ⓓ
136	Ⓐ	Ⓑ	Ⓒ	Ⓓ
137	Ⓐ	Ⓑ	Ⓒ	Ⓓ
138	Ⓐ	Ⓑ	Ⓒ	Ⓓ
139	Ⓐ	Ⓑ	Ⓒ	Ⓓ
140	Ⓐ	Ⓑ	Ⓒ	Ⓓ

3

Structural Depth Exam

The following set of questions numbered 201 to 240 is representative of a 4-hr structural depth exam according to the syllabus and guidelines for the Principles and Practice (P&P) of Civil Engineering Examination administered by the National Council of Examiners for Engineering and Surveying (NCEES), current for the October 2020 examination.

Enter your answers on page 74. Detailed solutions are on page 215.

201

A steel frame is shown in the following figure. The floor load tributary to each frame is 48,000 lb. The fundamental period (seconds) of vibration is most nearly:

A. 0.1

B. 0.2

C. 0.3

D. 0.4

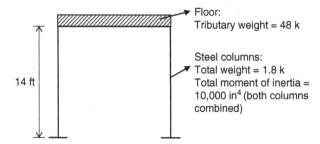

14 ft

Floor:
Tributary weight = 48 k

Steel columns:
Total weight = 1.8 k
Total moment of inertia =
10,000 in⁴ (both columns
combined)

202

A rectangular sawn lumber beam made of Southern Pine has a nominal 6 in × 12 in cross section (dressed dimensions 5.5 in × 11.25 in). The beam carries uniformly distributed loads $w_{DL} = 200$ lb/ft and $w_{LL} = 650$ lb/ft over a simple span $L = 24$ ft. The maximum shear stress (lb/in²) is most nearly:

A. 170

B. 250

C. 335

D. 495

203

A reinforced concrete beam ($f'_c = 5,000$ psi; $f_y = 60,000$ psi) has a rectangular section (width = 20 in, depth = 28 in) has a cracked moment of inertia $I_{cr} = 11,480$ in⁴ and a cracking moment $M_{cr} = 115$ k-ft. The beam carries a uniformly distributed factored load $w_u = 6$ k/ft over a simple span $L = 24$ ft. The maximum deflection (in) is most nearly:

A. 0.90

B. 1.20

C. 1.50

D. 1.80

204

A vertical load P acts on a propped cantilever beam as shown in the following figure. The beam section has the following properties—depth $d = 16$ in; strong axis moment of inertia $I_x = 1,240$ in^4, weak axis moment of inertia $I_y = 340$ in^4; plastic section moduli $Z_x = 189$ in^3; $Z_y = 46$ in^3. The steel used for the beam has the following properties: $F_y = 50$ ksi; $F_u = 65$ ksi.

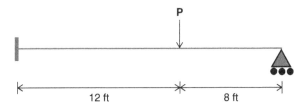

The maximum load P (k) that will create a plastic hinge collapse of the beam is most nearly:

A. 125
B. 160
C. 190
D. 230

205

A sawn rectangular timber beam (Douglas Fir, 6 in × 10 in nominal section) has a simple span L of 20 ft and supports a live load of 125 lb/ft and a dead load of 150 lb/ft. The maximum bending stress (psi) is most nearly:

A. 1,400
B. 1,600
C. 1,800
D. 2,000

206

A timber post serves as a compression member. The bottom support is adequately restrained against translation and rotation. The top of the column is braced in the weak direction, as shown. The cross section is 2 in × 6 in (nominal). The length of the column is 12 ft. Modulus of elasticity $E_{min} = 1.5 \times 10^6$ psi.

The Euler buckling load (k) is most nearly:

A. 1.7

B. 2.5

C. 3.4

D. 5.1

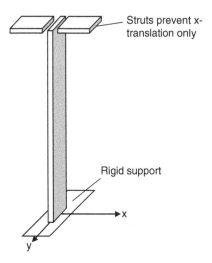

Struts prevent x-translation only

Rigid support

x

y

207

For the plane truss loaded as shown in the following figure, the force (k) in member CJ is most nearly:

A. 34 (tension)

B. 34 (compression)

C. 28 (tension)

D. 28 (compression)

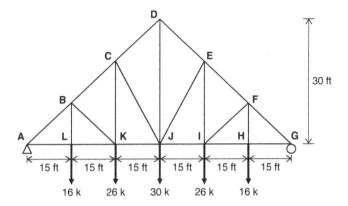

208

An elevated water tank is supported by four tower legs as shown in the following figure. The empty tank weighs 6 k and the full tank weighs 300 k. The resultant wind force of 120 k acts at a height of 65 ft as shown. Each tower leg is supported by an isolated square footing. The maximum design uplift force (k) for designing anchor bolts for each footing is most nearly:

A. 55.0
B. 128.5
C. 130.0
D. 258.5

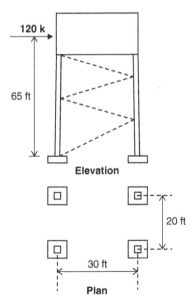

209

For the beam shown in the following figure, the maximum bending moment (k-ft) is most nearly:

A. 64
B. 72
C. 80
D. 88

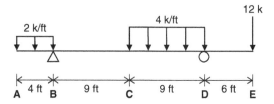

210

A two-member truss is loaded with a vertical load of 30 k at node B as shown in the following figure. The cross sections of the members AB and BC are: 2 in² for AB, 3 in² for BC. Modulus of elasticity $E = 29,000$ ksi for both members. The vertical deflection at B (in) is most nearly:

A. 0.05
B. 0.10
C. 0.15
D. 0.20

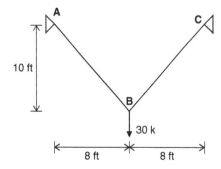

211

Computer analysis of a beam ABCD shown in the following figure results in the following joint moments:

$M_{BA} = 65.5$ k-ft (clockwise)
$M_{BC} = 65.5$ k-ft (counterclockwise)
$M_{CB} = 209.0$ k-ft (clockwise)
$M_{CD} = 209.0$ k-ft (counterclockwise)

The vertical reaction at B (k) is most nearly:

A. 12.8
B. 19.1
C. 26.4
D. 32.0

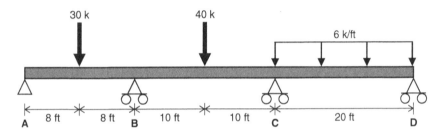

212

For the steel beam shown in the following figure, the following properties are given: $A = 24.7$ in^2; $I_x = 870$ in^4; $I_y = 215$ in^4). The vertical deflection (in) at point A is most nearly:

A. 0.005
B. 0.010
C. 0.017
D. 0.025

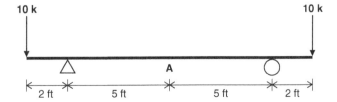

213

The propped cantilever beam shown carries a point load P. The beam is a W18 × 86 section. Assume $F_y = 50$ ksi; $F_u = 70$ ksi. The magnitude of the load P (k) that would cause collapse of the beam by plastic hinge formation is most nearly:

A. 210
B. 256
C. 388
D. 775

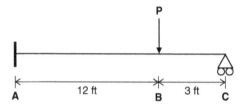

214

The unsymmetrical I-shaped section shown in the following figure is fabricated from A36 steel. The plastic moment capacity (k-ft) of the built-up section is most nearly:

A. 1,350
B. 1,650
C. 2,150
D. 2,650

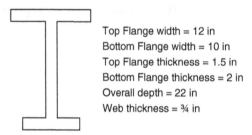

Top Flange width = 12 in
Bottom Flange width = 10 in
Top Flange thickness = 1.5 in
Bottom Flange thickness = 2 in
Overall depth = 22 in
Web thickness = ¾ in

215

A concrete bridge deck is exposed to freezing and thawing conditions with moisture and deicing chemical present continuously. Twenty-eight-day compression strength $f'_c = 6,000$ psi. Nominal maximum size of aggregate = 1 in. The minimum air content (%) is most nearly:

A. 2
B. 3
C. 4
D. 5

216

A prestressed concrete I-beam section is shown in the following figure. The neutral axis is located 13.73 in below the top fiber. Prestressing tendons (area = 5.90 in², ultimate stress $f_{pu} = 270$ ksi, initial prestress = $0.75f_{pu}$, prestress losses = 32 ksi) are located 12.2 in above the bottom edge of the beam. The flexural stress (lb/in²) on the top fiber due to a combination of the prestress force (after all losses) plus a bending moment (due to total DL + LL) = 1,400 k-ft acting on the beam (the gravity loads produce convex curvature on the bottom surface) is most nearly:

A. 845 (tension)
B. 1,630 (compression)
C. 2,145 (compression)
D. 1,230 (tension)

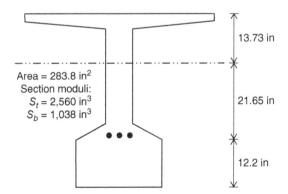

Area = 283.8 in²
Section moduli:
S_t = 2,560 in³
S_b = 1,038 in³

13.73 in

21.65 in

12.2 in

217

Concrete masonry units having a specified compressive strength of 1,800 psi are subject to a combination of axial and bending stress. The allowable compressive stress (lb/in²) is most nearly:

A. 450
B. 600
C. 720
D. 810

218

A three-story building has a rectangular plan (120 ft × 80 ft) and mean roof height = 42 ft. The fundamental period of the building is 0.35 sec. For the purposes of wind analysis and design according to the International Building Code, which of the following statements is true?

A. The building is considered low-rise and flexible.
B. The building is not considered rigid but not low-rise.
C. The building is considered flexible but not low-rise.
D. The building is considered low-rise but not flexible.

219

Shear reinforcement for a concrete beam is provided in the form of stirrups (no. 4 bars). The minimum inside radius (in) of the bend, according to provisions of ACI 318 is most nearly:

A. 1.0
B. 2.0
C. 3.0
D. 4.0

220

A circular reinforced concrete column supports a 300 k dead load and a 350 k live load. The compressive strength of the concrete is 4,000 psi and the yield stress of the reinforcement steel is 60 ksi. The longitudinal reinforcement is laterally confined by a ⅜-in-diameter spiral with a pitch of 3 in. If the maximum allowable reinforcement is used, the required diameter of the column (in) is most nearly:

A. 13
B. 14
C. 16
D. 17

221

What is the maximum shear (k) in a simple span bridge ($L = 125$ ft) due to AASHTO HL-93 loading (unfactored) occupying a single lane?

A. 65
B. 76
C. 107
D. 119

222

A W16 × 100 beam has a simple span of 30 ft and carries a uniformly distributed load. Steel grade is A992 grade 50. The compression flange of the beam has lateral support at the supports and at midspan only. Ignore C_b.

ASD	LRFD
The maximum load (k/ft) the beam can carry, as given by AISC-ASD provisions, is most nearly:	The maximum factored load (k/ft) the beam can carry, as given by AISC-LRFD provisions, is most nearly:
A. 3.3	A. 5.0
B. 4.0	B. 6.0
C. 4.7	C. 7.0
D. 5.2	D. 8.0

223

A 16 in × 20 in reinforced-concrete ($f_c' = 4{,}000$ psi, $f_y = 60{,}000$ psi) column is subjected to the following loads:

$P_D = 300$ k (concentric)
$P_L = 180$ k (eccentric; $e = 4$ in)

The required area of longitudinal reinforcement (in²) is most nearly:

A. 3.2
B. 3.8
C. 5.4
D. 6.0

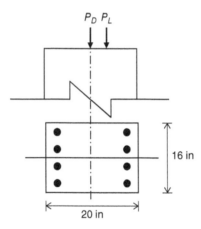

224

A36 steel ($F_y = 36$ ksi; $F_u = 58$ ksi) is used for a truss. All loads shown in the following figure are service loads. What is the minimum gross area (in²) needed for the bottom chord? Assume effective net area $= 0.75 \times A_g$. (Load is 30% DL + 70% LL).

A. 13.5
B. 17.5
C. 21.5
D. 25.5

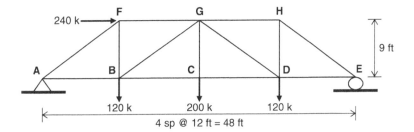

240 k→ F G H

9 ft

A B C D E

120 k   200 k   120 k

4 sp @ 12 ft = 48 ft

225

A reinforced-concrete $(f'_c = 4 \text{ ksi}; f_y = 60 \text{ ksi})$ beam is simply supported over a span $L = 25$ ft. The beam carries uniformly distributed loads $w_{DL} = 2$ k/ft and $w_{LL} = 4$ k/ft over its length. If the width of the beam is 15 in, the minimum satisfactory depth (in) is most nearly:

A. 23
B. 26
C. 29
D. 32

226

Determine if the given section is compact, non-compact, or slender. $F_y = 50$ ksi.

A. Compact web only
B. Non-compact flange only
C. Non-compact (as a whole section)
D. Compact (as a whole section)

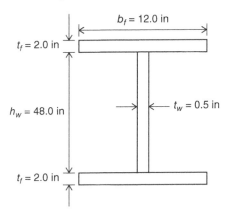

$b_f = 12.0$ in

$t_f = 2.0$ in

$h_w = 48.0$ in

$t_w = 0.5$ in

$t_f = 2.0$ in

227

A reinforced-concrete floor consists of a 5-in-thick slab cast monolithically with beams as shown. Twenty-eight-day compressive strength = 4,000 psi. Steel reinforcement is grade 60 deformed bars. Beams have simple span = 28 ft. The floor live load is 85 psf. Beams are to be designed with tension steel only. The required reinforcement (in²) of the tensile reinforcement is most nearly:

A. 2.8
B. 3.8
C. 4.8
D. 5.8

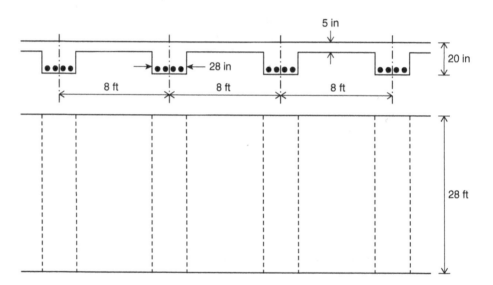

228

A steel column ($F_y = 36$ ksi) has a $KL_x = 40$ ft and $KL_y = 20$ ft. Find the smallest satisfactory W12 section that can support an axial load of 200 k dead load + 200 k live load. It is permissible to use either ASD or LRFD.

A. W12 × 79
B. W12 × 87
C. W12 × 96
D. W12 × 106

229

A 7.5 in × 16 in masonry lintel spans a door opening 5 ft wide. The wall above the lintel is 8 in thick with a unit weight of 130 lb/ft³. Assume 8-in bearing on either side of the opening. Assume unit weight of the reinforced lintel to be 140 lb/ft³. The top of the wall carries joists every 16 in. Each joist exerts a vertical reaction of 600 lb on the wall. The maximum bending moment (lb-ft) in the lintel is most nearly:

A. 810
B. 900
C. 990
D. 1080

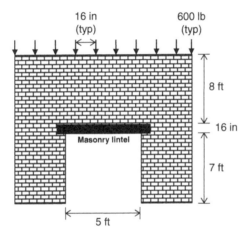

230

A composite concrete-deck steel-girder system for a simple span ($L = 70$ ft) bridge is shown. The beam is a doubly symmetric plate girder ($A_s = 136$ in^2; $I_{xx} = 81,940$ in^4; depth = 60 in). Slab thickness = 8 in.

Twenty-eight-day compressive strength of the concrete is 4 ksi. Yield stress for steel is 60 ksi. Load on the deck (including the weight of slab plus asphalt overlay and equivalent traffic load) may be taken as 1.8 k/ft^2. The maximum bending stress (k/in^2) in the steel girders is most nearly:

A. 32
B. 36
C. 39
D. 42

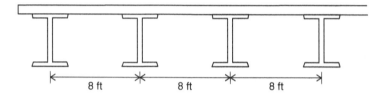

231

A C10 × 30 is used as a tension member as shown in the following figure. Steel used is A572 grade 50. Bolts used in the connection are ¾-in-diameter A325.

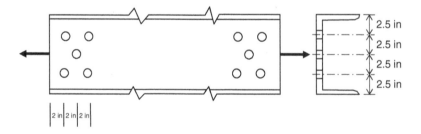

ASD	LRFD
The required strength (k) of the member, as given by AISC-ASD provisions, is most nearly:	The design strength (k) of the member, as given by AISC-LRFD provisions, is most nearly:

<table>
<tr><td>

A. 205
B. 240
C. 275
D. 305

</td><td>

A. 310
B. 360
C. 410
D. 460

</td></tr>
</table>

232

The cross section of a short, reinforced-concrete column is shown in the following figure. Assume that the eccentricity of the load is 4.0 in. Use $f'_c = 4,000$ psi; $f_y = 60,000$ psi. The design capacity (k) of the column is most nearly:

A. 360
B. 425
C. 465
D. 530

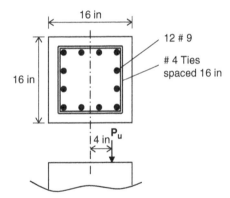

233

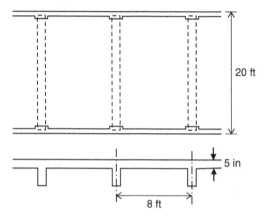

20 ft

5 in

8 ft

A floor system consists of a 5-in-thick reinforced-concrete slab built monolithically with concrete beams spaced 8 ft as shown. The beams span 20 ft between simple supports. Use $f'_c = 4,500$ psi and $f_y = 60,000$ psi. The service loads on the slab are: Superimposed dead load = 40 lb/ft². Live load = 85 lb/ft² in the slab. The flexural reinforcement required (in²/ft) at the critical section for positive moment is most nearly:

A. 0.32
B. 0.21
C. 0.14
D. 0.11

234

A steel tension member has a W12 × 72 section and is connected through its flanges using ¾-in-diameter high-strength A325 bolts as shown. $F_y = 36$ ksi; $F_u = 58$ ksi. The nominal strength (k) based on block shear is most nearly:

A. 550
B. 1,090
C. 820
D. 275

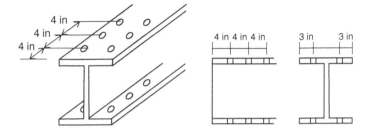

235

For the design of main wind-force resisting system in a building, what is the minimum percentage of the gross wall area that wall openings must occupy in order for a building to be considered an open building?

A. 80
B. 75
C. 90
D. 85

236

Which of the following statements must be true in order to determine the design wind pressures based on the simplified procedure in ASCE 7?

    I. The mean roof height h must be less than or equal to 60 ft ($h \le 60$ ft).
    II. The building has response characteristics making it subject to vortex shedding.
    III. The building is not classified as a flexible building.
    IV. The building does not have response characteristics making it subject to instability due to galloping or flutter.

A. I
B. I and II
C. I, III, and IV
D. All of the above

237

According to the AISC Manual of Steel Construction, the preferred ASTM grade for square HSS shapes is

A. A500
B. A242
C. A992
D. A36

238

According to the 29 CFR 1910, the preferred pitch angle (with the horizontal plane) of fixed ladders is:

A. 50°–60°
B. 55°–65°
C. 60°–75°
D. 66°–75°

239

A bridge consists of a concrete slab supported by four equally spaced girders as shown in the following figure. The concrete deck is 36 ft wide and 9 in thick. Steel plate girders weigh 282 lb/ft. Each girder is simply supported over a span $L = 80$ ft. Skew angle $= 0°$. Repair of piers requires that a temporary jacking tower be located adjacent to each pier and a line of six hydraulic jacks be used to raise the girders using a jacking beam.

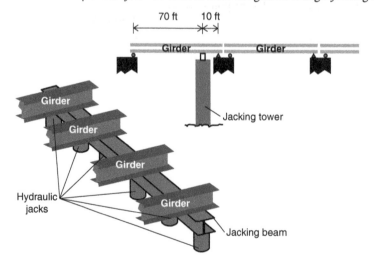

Using a factor of safety of 2.0, the minimum required capacity (tons) of the hydraulic jacks is most nearly:

A. 45
B. 40
C. 35
D. 30

240

A structure has been examined to determine a need for rehabilitation. The related costs are summarized as follows:

Current annual costs = $40,000
Estimated rehabilitation cost = $350,000
Annual costs projected after rehabilitation = $15,000
Expected useful life remaining = 20 years
Projected increase in residual value (at end of useful life) = $200,000

The return on investment (ROI) for performing the rehabilitation is most nearly:

A. 5%
B. 6%
C. 7%
D. 8%

END OF STRUCTURAL DEPTH EXAM

USE THE ANSWER SHEET ON THE NEXT PAGE

Structural Depth Exam: Answer Sheet

201	Ⓐ	Ⓑ	Ⓒ	Ⓓ
202	Ⓐ	Ⓑ	Ⓒ	Ⓓ
203	Ⓐ	Ⓑ	Ⓒ	Ⓓ
204	Ⓐ	Ⓑ	Ⓒ	Ⓓ
205	Ⓐ	Ⓑ	Ⓒ	Ⓓ
206	Ⓐ	Ⓑ	Ⓒ	Ⓓ
207	Ⓐ	Ⓑ	Ⓒ	Ⓓ
208	Ⓐ	Ⓑ	Ⓒ	Ⓓ
209	Ⓐ	Ⓑ	Ⓒ	Ⓓ
210	Ⓐ	Ⓑ	Ⓒ	Ⓓ
211	Ⓐ	Ⓑ	Ⓒ	Ⓓ
212	Ⓐ	Ⓑ	Ⓒ	Ⓓ
213	Ⓐ	Ⓑ	Ⓒ	Ⓓ
214	Ⓐ	Ⓑ	Ⓒ	Ⓓ
215	Ⓐ	Ⓑ	Ⓒ	Ⓓ
216	Ⓐ	Ⓑ	Ⓒ	Ⓓ
217	Ⓐ	Ⓑ	Ⓒ	Ⓓ
218	Ⓐ	Ⓑ	Ⓒ	Ⓓ
219	Ⓐ	Ⓑ	Ⓒ	Ⓓ
220	Ⓐ	Ⓑ	Ⓒ	Ⓓ

221	Ⓐ	Ⓑ	Ⓒ	Ⓓ
222	Ⓐ	Ⓑ	Ⓒ	Ⓓ
223	Ⓐ	Ⓑ	Ⓒ	Ⓓ
224	Ⓐ	Ⓑ	Ⓒ	Ⓓ
225	Ⓐ	Ⓑ	Ⓒ	Ⓓ
226	Ⓐ	Ⓑ	Ⓒ	Ⓓ
227	Ⓐ	Ⓑ	Ⓒ	Ⓓ
228	Ⓐ	Ⓑ	Ⓒ	Ⓓ
229	Ⓐ	Ⓑ	Ⓒ	Ⓓ
230	Ⓐ	Ⓑ	Ⓒ	Ⓓ
231	Ⓐ	Ⓑ	Ⓒ	Ⓓ
232	Ⓐ	Ⓑ	Ⓒ	Ⓓ
233	Ⓐ	Ⓑ	Ⓒ	Ⓓ
234	Ⓐ	Ⓑ	Ⓒ	Ⓓ
235	Ⓐ	Ⓑ	Ⓒ	Ⓓ
236	Ⓐ	Ⓑ	Ⓒ	Ⓓ
237	Ⓐ	Ⓑ	Ⓒ	Ⓓ
238	Ⓐ	Ⓑ	Ⓒ	Ⓓ
239	Ⓐ	Ⓑ	Ⓒ	Ⓓ
240	Ⓐ	Ⓑ	Ⓒ	Ⓓ

4

Geotechnical Depth Exam

The following set of questions numbered 301 to 340 is representative of a 4-hr geotechnical depth exam according to the syllabus and guidelines for the Principles and Practice (P&P) of Civil Engineering Examination administered by the National Council of Examiners for Engineering and Surveying (NCEES), current for the October 2020 examination.

Enter your answers on page 100. Detailed solutions are on page 239.

301

Results from a sand cone test are listed below:

Net weight of soil obtained from the test hole:	5.52 lb
Moisture content of soil from test hole:	19%
Unit weight of dry test sand:	88.2 lb/ft^3
Initial weight of sand cone apparatus filled with test sand:	13.75 lb
Final weight of sand cone apparatus after sand fills test hole:	10.24 lb

A standard Proctor test conducted on the soil resulted in:

Maximum dry unit weight	126.3 lb/ft^3
Optimum moisture content	17.5%

The in-place percent compaction is most nearly:

A. 92

B. 97

C. 114

D. 133

302

The following data is given for a soil sample:

Sieve Analysis:

Sieve Size	Percent Retained
No. 4	8
No. 10	10
No. 20	12
No. 40	21
No. 100	15
No. 200	8

Atterberg Tests:

Liquid limit = 43
Plastic limit = 21

The USCS soil classification is:

A. SM

B. GW

C. GP

D. SC

303

A 20-ft rock core is shown in the following figure. The RQD is most nearly:

A. 97%

B. 93%

C. 85%

D. 89%

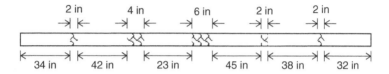

304

Which of the following statements is false?

A. Ground penetrating radar works better for clay soils than sands.

B. Seismic reflection methods are little used for shallow ground investigation.

C. The primary limitation of conducting a gravity survey is high equipment cost.

D. Resistivity surveys are not very effective in site investigation.

305

A cone penetrometer test produces a continuous record of tip and sleeve resistance as well as pore pressure induced behind the cone tip. At a certain depth, the following are recorded:

Tip resistance = 5.2 tons/ft²
Sleeve resistance = 0.03 tons/ft²
Pore pressure = 0.9 tons/ft²

The soil is most likely a:

A. sand

B. silt

C. clay

D. peat

306

A sample of wet soil weighs 1,331.5 grams. The sample is then coated with wax (specific gravity = 0.9) and then weighed to be 1,368.2 g. The wax-coated sample is then completely immersed in water and found to weigh 593.4 g. Water content of the soil sample is 15.2% and specific gravity of solids is 2.70. The degree of saturation (%) of the soil sample is most nearly:

A. 57
B. 52
C. 45
D. 37

307

Three cylindrical samples (diameter = 4 in, thickness = 2 in) of the same soil are tested in direct shear using a split mold and subjected to a shear force pair as shown in the figure. The top and bottom surfaces of the sample are capped with porous stone loading plates. The normal force N (lb) is varied for each sample. The horizontal shear force F (lb) required to cause shear failure is also recorded. These values are summarized in the table. The cohesion of the soil (lb/ft²) is most nearly:

A. 370
B. 575
C. 760
D. 980

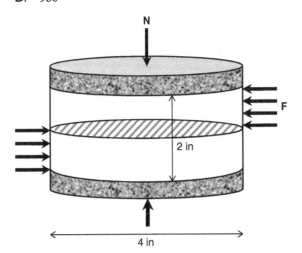

Sample	N (lb)	F (lb)
1	120	78
2	160	93
3	220	116

308

The soil profile shown in the following figure consists of a silty sand layer overlying a 12-ft-thick layer of medium sand. A standard penetration test is conducted with a split spoon sampler, which is driven through an 18-in penetration from depth of 10 ft to 11 ft 6 in. The number of blows to drive the sampler through 6-in-penetration intervals are 12, 17, and 22, respectively.

The corrected standard penetration resistance (N-value) is most nearly:

A. 31
B. 39
C. 57
D. 62

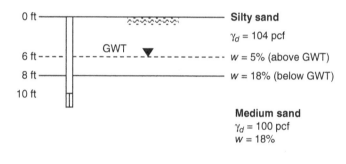

309

A dam of length 120 ft is constructed of impermeable soil. The dam overlays a sand bed of thickness 6.5 ft as shown in the following figure. The sand has the following characteristics:

Unit weight = 125 lb/ft³
Void ratio = 0.45
Water content = 18%
Hydraulic conductivity = 1×10^{-4} ft/sec
Scour velocity for sand = 8 in/hr

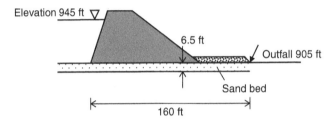

The factor of safety for scouring in the sand drain is most nearly:

A. 0.9
B. 1.8
C. 2.3
D. 2.9

310

Undrained triaxial tests are conducted on a sample of silty clay. The following results are obtained:

Sample diameter = 2 in
Sample length = 4 in
Radial (hydrostatic) stress maintained during test = 18 psi
Added axial load at failure = 158.2 lb
Pore pressure at failure = 5.6 psi

If an identical sample is tested in a drained test in which the radial stress is increased to 36 psi, the expected axial load at failure (lb) is most nearly:

A. 290
B. 240
C. 190
D. 160

311

Results from a standard Proctor compaction test with six soil samples from a borrow pit are tabulated in the following table. The natural moisture content of the excavated material is 12%. The fill location requires 1.5 million yd³ of soil compacted to a minimum 90% of the maximum Proctor dry density.

Sample	Net Weight of Soil (lb)	Moisture Content (%)
1	3.24	12
2	3.70	14
3	3.95	16
4	4.21	18
5	3.90	20
6	3.40	22

The total volume of borrow soil that must be excavated (yd³) is most nearly:

A. 1.72 million

B. 1.65 million

C. 1.53 million

D. 1.42 million

312

To construct an embankment it will take 500,000 yd³ of soil compacted to at least 95% of the standard Proctor maximum dry density (MDD). In a proposed borrow pit the soil has an in-situ total density $\gamma = 108$ pcf at a water content of 20%. It is planned to transport the soil using trucks having a 10 yd³ capacity of the borrow soil at a total density of 115 pcf and water content of 20%. If the standard Proctor MDD for this soil equals 95 pcf and the optimum moisture content equals 23%, the number of truck loads required to construct the embankment assuming no loss of soil is most nearly:

A. 42,000 truck loads

B. 47,000 truck loads

C. 52,000 truck loads

D. 57,000 truck loads

313

A retaining wall provides lateral support to a granular backfill ($\phi = 34°$) to a height of 16 ft as shown in the following figure. The backfill surface is horizontal and is used to stockpile material excavated from the site.

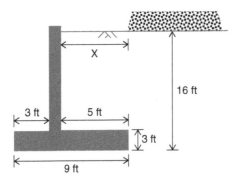

The minimum setback distance X (ft) is most nearly:

A. 8.5
B. 11.9
C. 13.5
D. 15.0

314

For a site, the design earthquake has the following parameters:

Recurrence interval = 540 years
Richter's magnitude = 7.2
Cyclic stress ratio = 0.23

The surface soil layer is a 30-ft-deep bed of fine sand with the following characteristics:

Unit weight $\gamma = 124$ lb/ft³
Angle of internal friction = 34°
Relative density = 0.85
Depth of water table = 10 ft

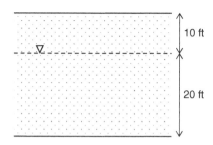

The following results are produced by a laboratory test on a similar soil sample:

Relative density = 95%
Shear stress to cause liquefaction = 1,200 lb/ft^2

The factor of safety for liquefaction, at a depth of 30 ft, is most nearly:

A. 1.6
B. 1.9
C. 2.2
D. 2.5

315

Subsurface exploration at a site shows the data summarized in the following table:

Depth below Surface (ft)	Description	SPT N (blows/ft)
0–12	Silty sand	12
12–32	Dense sand	34
32–65	Brown silty clay	25
65–88	Dense clayey silt	33
88–125	Clay	22
125–150		45

The seismic site class for this location is:

A. A
B. B
C. C
D. D

316

A planned excavation is to be 15-ft deep. The soil will be subject to vibration from pile driving operations. The unconfined compressive strength of the soil is 3,250 psf. According to the guidelines of the OSHA 1926 regulations (Title 29 CFR), the classification of the soil is:

A. Type C
B. Type A
C. Type D
D. Type B

317

A temporary slope will be excavated in clay as shown in the following figure. The height of the embankment is 20 ft and the slope angle is 30°. A presumed failure surface (base circle) which is the tangent to a firm layer is shown (dashed line). The firm layer is at a depth of 30 ft below the base of the slope. For this failure surface, the factor of safety for slope stability is most nearly:

A. 1.5
B. 1.8
C. 2.1
D. 2.4

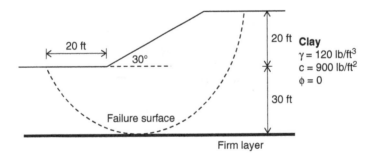

318

A flexible pavement consists of three layers as shown in the figure below.

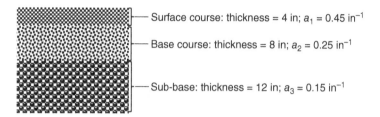

— Surface course: thickness = 4 in; $a_1 = 0.45$ in^{-1}

— Base course: thickness = 8 in; $a_2 = 0.25$ in^{-1}

— Sub-base: thickness = 12 in; $a_3 = 0.15$ in^{-1}

The structural number (SN) is most nearly:

A. 4.5
B. 5.5
C. 6.5
D. 7.5

319

The soil profile shown in the following figure consists of a sandy fill layer overlying a layer of normally consolidated clay with the soil properties shown. The soil below the elevation of −20 ft is stiff clay.

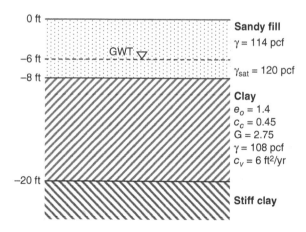

(*Continued on next page*)

The GWT is lowered to the top of the clay layer and a mat foundation (bottom of mat at depth of 42 in below ground) of plan dimensions 190 ft × 250 ft is constructed. Total vertical (concentric) load on the mat = 20,000 tons. The time (years) for 80% consolidation to occur is most nearly:

A. 6
B. 9
C. 14
D. 18

320

A 30-ft-deep aquifer is confined between two impermeable layers of rock as shown. The elevation of the piezometric surface is at 20 ft above the top of the aquifer. A 9-in-diameter well is used to establish a steady state pumping rate of 2,000 gal/min. The hydraulic conductivity of the soil in the aquifer is 1,000 ft/day. Observation wells 1 and 2 are drilled at a radial distance of 30 ft and 180 ft from the centerline of the pumping well as shown in the following figure. The drawdown of the piezometric surface at observation well no. 1 is 4.5 ft. The drawdown of the piezometric surface (ft) at observation well no. 2 is most nearly:

A. 2.1
B. 2.8
C. 3.2
D. 3.8

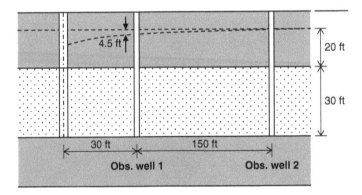

321

A reinforced-concrete dam has a sheet pile cutoff wall parallel to its upstream face as shown. The elevation of the reservoir is 213.45 ft A.S.L. The underlying soil is silty sand with hydraulic conductivity K = 200 ft/day. The dam is 150 ft long. The flow net has been constructed and is shown. The total seepage loss (ft³/sec) under the dam is most nearly:

A. 0.013
B. 0.1
C. 2.0
D. 13.7

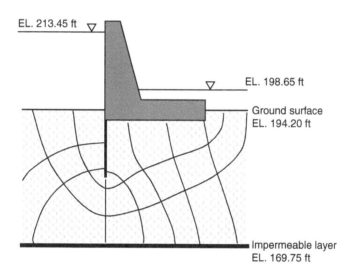

322

Water is allowed to percolate through a sand filter as shown in the following figure. The hydraulic conductivity of the sand is 0.012 ft/hr. If the elevation of the water column in the fine bore tube (diameter 0.1 in) at $t = 0$ is 238.56 ft above sea level, the elevation (ft) at $t = 10$ min is most nearly:

A.   232.02
B.   232.80
C.   233.18
D.   233.74

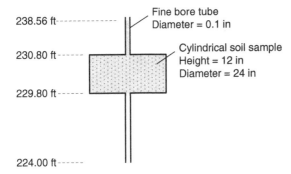

323

Which of the following statements is/are true?

    I.   Karst refers to a type of soil created from volcanic ash.
    II. Karst formations create acidity in surface runoff.
    III. Karst formations form soil deposits that have high bearing capacity.
    IV. Karst formations contribute to subsurface storage of water.

A.   II and IV only
B.   II, III, and IV
C.   I and III only
D.   All of them

324

The corrosion potential of soils depends on its electrical resistivity. Which of the following statements is true?

A. Cast iron pipes buried in sandy soils have low corrosion potential because of low electrical resistivity of the soil.

B. Cast iron pipes buried in clay soils have low corrosion potential because of low electrical resistivity of the soil.

C. Cast iron pipes buried in sandy soils have high corrosion potential because of low electrical resistivity of the soil.

D. Cast iron pipes buried in clay soils have high corrosion potential because of low electrical resistivity of the soil.

325

Which of the following factors are significant in determining the frost heave in soils?

    I. Grain size
   II. Unit weight
  III. Moisture content
  IV. Temperature fluctuations
   V. Overburden pressure

A. All of them
B. All except I
C. All except II
D. All except V

326

A trench is excavated in sandy soil to a depth of 20 ft. The trench is supported by timber planking as shown. The horizontal earth pressure behind the sheet piles is calculated as 500 lb/ft². The vertical sheet piles are supported by longitudinal wales (vertical spacing = 4 ft) as shown. If the allowable bending stress in the timber is 1,400 lb/in², the required thickness (in) of the planks is most nearly:

A.   1.0
B.   1.5
C.   1.75
D.   2.0

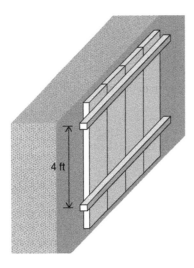

4 ft

327

A 20-ft-wide × 24-ft-deep trench in a clayey soil is braced as shown in the following figure. Longitudinal spacing of the struts is 8 ft. The axial load in strut S_1 (k) is most nearly:

A.   10
B.   15
C.   20
D.   25

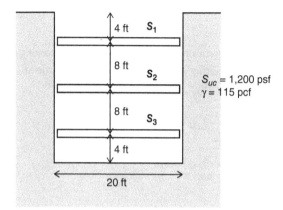

328

A reinforced-concrete cantilever retaining wall is shown in the following figure. The total width of the wall footing is 9 ft. Excluding the effect of the passive soil, the factor of safety against overturning is most nearly:

A. 3.7

B. 4.7

C. 2.1

D. 3.1

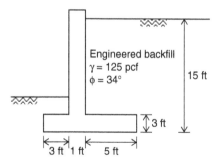

329

A cantilever reinforced-concrete retaining wall is shown in the following figure. The friction angle between the wall footing and the soil is 20°. The factor of safety for overturning of the wall is most nearly:

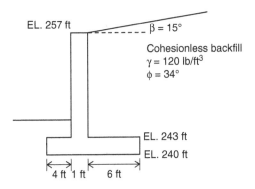

EL. 257 ft

β = 15°

Cohesionless backfill
$\gamma = 120 \text{ lb/ft}^3$
$\phi = 34°$

EL. 243 ft
EL. 240 ft

4 ft 1 ft   6 ft

A. 1.2
B. 2.5
C. 3.4
D. 4.8

330

An anchored bulkhead is used to retain soil behind a vertical trench as shown in the following figure. The dredge side elevation is 120 ft and water table is at the elevation of 134 ft. The anchor is located 1 ft above the water table. The bulkhead is driven to a depth of 105 ft. The total lateral pressure (lb/ft) on the bulkhead due to the active soil is most nearly:

A. 25,200
B. 31,400
C. 41,300
D. 48,700

140 ft

134 ft ▽

γ_{sat} = 123 lb/ft^3
ϕ = 32°
e = 0.7
G_s = 2.65

120 ft

105 ft

331

A wall footing is embedded 30 in. in a sand layer as shown in the following figure. Footing width = 4 ft. Superstructure loads are:

   Concentric load = 10 k/ft
   Moment = 9 k-ft/ft

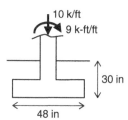

10 k/ft
9 k-ft/ft
30 in
48 in

The maximum soil pressure (lb/ft²) under the footing is most nearly:

A. 4,750
B. 5,875
C. 6,000
D. 6,750

332

A 10-ft square footing embedded 2 ft into a sand profile has a total density, γ_t, equal to 115 pcf and bearing capacity factors $N_c = 48$, $N_q = 25$, and $N_\gamma = 19$. Neglecting shape factors, if the depth to the water table is 7 ft below the ground surface and the proposed column load equals 750 k, the FS against bearing capacity failure is most nearly:

A. 1.5
B. 1.8
C. 2.2
D. 2.6

333

A 5-ft × 5-ft square footing carries a concentric load of 100 k. The depth of the footing is 3 ft. The vertical stress increase (lb/ft²) at a point 4 ft below and 6 ft laterally offset from the center of the footing is most nearly:

A. 100
B. 150
C. 200
D. 250

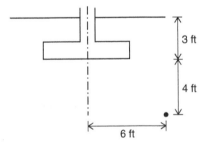

334

The square footing shown in the following figure supports a concentric column load. Minimum factor of safety based on ultimate bearing capacity is 2.8. The maximum column load (k) is most nearly:

A. 320
B. 290
C. 260
D. 230

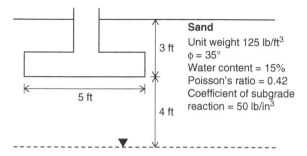

Sand
Unit weight 125 lb/ft³
$\phi = 35°$
Water content = 15%
Poisson's ratio = 0.42
Coefficient of subgrade
reaction = 50 lb/in³

3 ft

4 ft

5 ft

335

A rectangular combined footing supports two columns (marked A and B) as shown in the following figure. The underlying soil is medium sand ($\phi = 32°$, $\gamma = 127$ pcf). The concentric loads on columns A and B are:

$P_A = 100$ k
$P_B = 120$ k

The maximum soil pressure (lb/ft²) under the footing is most nearly:

A. 3,100
B. 3,600
C. 4,000
D. 4,300

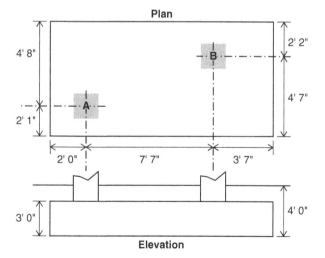

Plan

Elevation

336

A four-story building is supported by a group of vertical piles as shown in the following figure. The plan dimensions of the pile cap are 40 ft × 100 ft. Floor height is 15 ft. Based on the pile dimension and the type of soil adjacent to the pile, the following (ultimate) capacities have been established:

  Ultimate point bearing capacity = 70 k
  Ultimate side friction capacity = 40 k
  Weight of each pile = 5.6 k

Loads:

  Dead loads: Floors 1–4: 150 k
         Roof: 60 k
  Live loads: Floors 1–4: 220 k
         Roof: 70 k
  Lateral loads (wind) are shown on elevation.

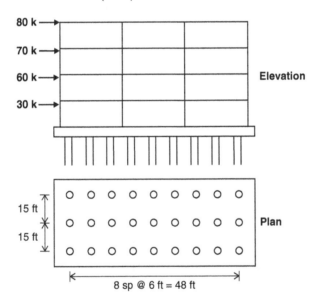

For the case shown (wind parallel to the long plan dimension), the maximum net uplift (k) at a pile is most nearly:

A. 8

B. 13

C. 24

D. 38

337

A pile group is as shown in the following figure (all dimensions are between pile centers). Piles are driven precast concrete piles of diameter 24 in. The total concentric superstructure load (including the weight of the pile cap) is 125 tons. The settlement (in) due to the consolidation of the clay layer is most nearly:

A. 1.3

B. 2.5

C. 3.8

D. 5.0

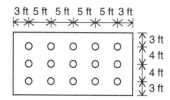

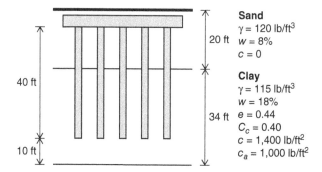

338

The footing for a bridge is supported by a group of eight precast concrete point-bearing piles as shown in the following figure. Each pile has an allowable load of 30 tons. Service loads from the superstructure transmitted to the pile group are:

Axial load: DL + LL (including weight of pile cap) = 350 k
Moment: DL + LL = 300 ft-k
28-day compression strength of concrete f'_c = 4,000 psi
Yield strength of reinforcement steel f_y = 60,000 psi

(*Continued on next page*)

The minimum required spacing B (ft) of the piles is most nearly:

A. 2 ft 6 in

B. 3 ft 0 in

C. 3 ft 6 in

D. 4 ft 0 in

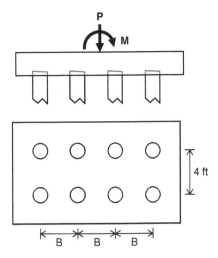

339

It is proposed to drive a concrete pile using a pile hammer with energy of 50,000 ft-lb. The minimum factor of safety against bearing capacity failure is equal to 6.0. The superstructure load transmitted to the pile is expected to be 40 tons.

The ENR pile driving equation gives the static (ultimate) bearing capacity of a pile based on a pile driving test as:

$$Q_{ult} = \frac{WH}{S + 1.0},$$

where Q_{ult} = ultimate capacity (lb)

W = weight of the ram (lb)

H = height the ram falls (in)

S = pile set in in. per blow

The required blow count (blows per ft) to achieve a design capacity of 40 tons is most nearly:

A. 12
B. 28
C. 35
D. 50

340

A structure has been examined to determine a need for rehabilitation. The related costs are summarized as follows:

Current annual costs = $40,000
Estimated rehabilitation cost = $350,000
Annual costs projected after rehabilitation = $15,000
Expected useful life remaining = 20 years
Projected increase in residual value (at end of useful life) = $200,000

The return on investment (ROI) for performing the rehabilitation is most nearly:

A. 5%
B. 6%
C. 7%
D. 8%

END OF GEOTECHNICAL DEPTH EXAM

USE THE ANSWER SHEET ON THE NEXT PAGE

Geotechnical PM Exam: Answer Sheet

301	Ⓐ	Ⓑ	Ⓒ	Ⓓ
302	Ⓐ	Ⓑ	Ⓒ	Ⓓ
303	Ⓐ	Ⓑ	Ⓒ	Ⓓ
304	Ⓐ	Ⓑ	Ⓒ	Ⓓ
305	Ⓐ	Ⓑ	Ⓒ	Ⓓ
306	Ⓐ	Ⓑ	Ⓒ	Ⓓ
307	Ⓐ	Ⓑ	Ⓒ	Ⓓ
308	Ⓐ	Ⓑ	Ⓒ	Ⓓ
309	Ⓐ	Ⓑ	Ⓒ	Ⓓ
310	Ⓐ	Ⓑ	Ⓒ	Ⓓ
311	Ⓐ	Ⓑ	Ⓒ	Ⓓ
312	Ⓐ	Ⓑ	Ⓒ	Ⓓ
313	Ⓐ	Ⓑ	Ⓒ	Ⓓ
314	Ⓐ	Ⓑ	Ⓒ	Ⓓ
315	Ⓐ	Ⓑ	Ⓒ	Ⓓ
316	Ⓐ	Ⓑ	Ⓒ	Ⓓ
317	Ⓐ	Ⓑ	Ⓒ	Ⓓ
318	Ⓐ	Ⓑ	Ⓒ	Ⓓ
319	Ⓐ	Ⓑ	Ⓒ	Ⓓ
320	Ⓐ	Ⓑ	Ⓒ	Ⓓ

321	Ⓐ	Ⓑ	Ⓒ	Ⓓ
322	Ⓐ	Ⓑ	Ⓒ	Ⓓ
323	Ⓐ	Ⓑ	Ⓒ	Ⓓ
324	Ⓐ	Ⓑ	Ⓒ	Ⓓ
325	Ⓐ	Ⓑ	Ⓒ	Ⓓ
326	Ⓐ	Ⓑ	Ⓒ	Ⓓ
327	Ⓐ	Ⓑ	Ⓒ	Ⓓ
328	Ⓐ	Ⓑ	Ⓒ	Ⓓ
329	Ⓐ	Ⓑ	Ⓒ	Ⓓ
330	Ⓐ	Ⓑ	Ⓒ	Ⓓ
331	Ⓐ	Ⓑ	Ⓒ	Ⓓ
332	Ⓐ	Ⓑ	Ⓒ	Ⓓ
333	Ⓐ	Ⓑ	Ⓒ	Ⓓ
334	Ⓐ	Ⓑ	Ⓒ	Ⓓ
335	Ⓐ	Ⓑ	Ⓒ	Ⓓ
336	Ⓐ	Ⓑ	Ⓒ	Ⓓ
337	Ⓐ	Ⓑ	Ⓒ	Ⓓ
338	Ⓐ	Ⓑ	Ⓒ	Ⓓ
339	Ⓐ	Ⓑ	Ⓒ	Ⓓ
340	Ⓐ	Ⓑ	Ⓒ	Ⓓ

5

Water Resources & Environmental Depth Exam

The following set of questions numbered 401 to 440 is representative of a 4-hr Water Resources & Environmental Depth exam according to the syllabus and guidelines for the Principles and Practice (P&P) of Civil Engineering Examination administered by the National Council of Examiners for Engineering and Surveying (NCEES), current for the October 2020 examination.

Enter your answers on page 126. Detailed solutions are on page 261.

401

A municipal plant receives water with a total hardness of 200 mg/L. The designed discharge hardness is 50 mg/L. An ion exchange unit with an overall efficiency of 88% is used for hardness reduction. The bypass factor (%) is most nearly:

A. 10
B. 15
C. 20
D. 25

402

The schematic of a wastewater treatment plant is shown in the following figure.

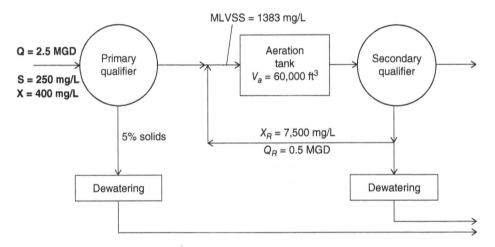

The total suspended solids (TSS) for the influent into the primary clarifier is 400 mg/L. The primary clarifier reduces total suspended solids by 60% and BOD by 15%. The volume of primary sludge produced per day (gal/day) is most nearly:

A. 125,000
B. 65,000
C. 12,000
D. 8,500

403

The flow rate treated at a wastewater treatment plant is 3.2 MGD. The TSS concentration in the influent is 800 mg/L. The flow passes through a bank of filters, arranged in parallel.

Filter size is limited to 250 ft².
The maximum solids load on each filter is 15 lb-TSS/ft²-day.
The maximum surface loading velocity on the filters = 10 ft/hr.

The number of filters needed is most nearly:

A. 6
B. 7
C. 8
D. 9

404

An orifice meter (orifice diameter = 2 in) is inserted in a 4-in-diameter cast iron pipe carrying water at 15°C. The pressure difference across the meter is 30 lb/in². The orifice has the following coefficients: $C_c = 0.9$, $C_v = 0.95$. The discharge (gal/min) through the pipe is most nearly:

A. 420
B. 570
C. 760
D. 920

405

Water flows under pressure in a horizontal circular pipe (diameter 8 in). The flow rate is 650 gal/min. The flow goes through a 2:1 reducer. The pressure loss across the reducer is 12 psi. The total head loss (ft) across the reducer is most nearly:

A. 12
B. 16
C. 20
D. 24

406

For a 12-in-diameter pipe ($C = 110$) conveying water at a flow rate of 1,500 gal/min, the pressure loss due to friction (psi per ft) is most nearly:

A. 0.0021
B. 0.0032
C. 0.0045
D. 0.0072

407

A nozzle (0.5 in diameter, $C_v = 0.90$, $C_d = 0.81$) discharges freely from the bottom of a cylindrical tank as shown in the following figure. The discharge (gal/min) from the nozzle is most nearly:

A. 7
B. 9
C. 12
D. 15

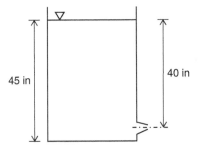

408

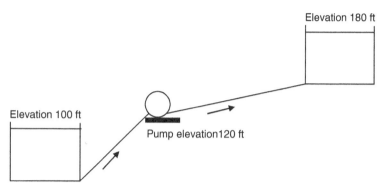

Elevation 180 ft

Elevation 100 ft

Pump elevation120 ft

The pump located at an elevation of 120 ft is used to pump a flow $Q = 3,000$ gpm from a reservoir with a surface elevation of 100 ft to another reservoir with a surface elevation of 180 ft. The approximate efficiency of the pump is 88%. The characteristics of the pipe system are as follows:

Suction line:	800 ft length, 18 in diameter, friction factor = 0.024
	Total minor loss coefficient (valves, bends, etc.) = 5
Discharge line:	2,500 ft length, 12 in diameter, friction factor = 0.026
	Total minor loss coefficient (valves, bends, etc.) = 25

The required brake horsepower of the pump is most nearly:

A.   100
B.   130
C.   160
D.   200

409

The pipe system shown in the following figure circulates water at 65°F. Flow enters the network at B and leaves it at node G. The table shows hydraulic characteristics of all pipes in the network. When the flow rate in and out of the network is 300 gal/min, the total head loss between nodes A and H is 70 ft.

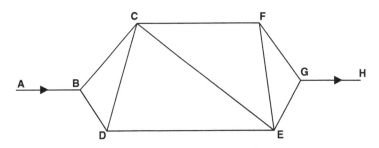

Segment	Length (ft)	Diameter (in)	Friction Factor
AB	250	8	0.020
BC	450	12	0.020
BD	160	8	0.025
CD	500	12	0.030
DE	800	12	0.030
CF	600	12	0.020
FE	500	12	0.025
CE	800	12	0.030
FG	500	12	0.020
EG	200	8	0.030
GH	300	8	0.025

If the flow rate (incident at A) changes to 500 gal/min, the pressure loss (lb/in²) between nodes A and H is most nearly:

A. 65

B. 100

C. 50

D. 80

410

A trapezoidal open channel with bottom width 10 ft and side slopes 2H:1V conveys a flow rate of 150 ft³/s. If the Manning's n is 0.016 and the longitudinal slope of the channel bottom is 0.4%, the depth of flow (in) is most nearly:

A. 15
B. 33
C. 29
D. 20

411

A rectangular channel is 15 ft wide and conveys a flow rate of 120,000 gal/min. The flow is conveyed down a spillway of the same width, at the bottom of which the depth of flow is 15 in. If a hydraulic jump is to be forced at the bottom of the spillway, the required tailwater depth (ft) is most nearly:

A. 2.0
B. 3.4
C. 5.2
D. 6.2

412

A 36-in-diameter reinforced-concrete pipe under a roadway serves as a culvert operating under inlet control. The following data apply to the culvert:

    Upstream invert: elevation 325.0
    Downstream invert: elevation 322.5
    Length of culvert: 145 ft
    Entrance condition: headwall with square edge
    Roadway surface elevation: 337.8 ft
    Culvert peak flow rate due to 100-year flood event: 100 cfs

(Continued on next page)

The minimum vertical clearance (ft) between the roadway and the 100-year flood elevation is most nearly:

A. 5.2

B. 2.2

C. 4.8

D. 7.8

413

Water flows through a rectangular open channel at a constant flow rate. At a particular location of the channel, the floor of the channel is raised 4 in for a 12-in-long step, as shown in the following figure. The depth over the step is measured as 9 in. The flow rate per unit width of the channel (cfs/ft) is most nearly:

A. 1.57

B. 1.72

C. 1.85

D. 1.98

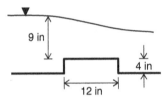

414

A rectangular open channel of width = 10 ft has a critical velocity = 15 ft/s. The flow rate (ft^3/s) at this critical velocity is most nearly:

A. 800

B. 1,000

C. 1,200

D. 1,400

415

A 280-acre development is under construction, which is estimated to last 3 years. During construction, the sediment yield is 1,200 ft³/acre-year. Following the completion of construction, the sediment yield is estimated to be 300 ft³/acre-year. A 5-acre water quality pond is used for sediment control. The bottom elevation of the pond is 345.6 ft above sea level and pool surface elevation is regulated at 353.4 ft above sea level. The minimum depth of water in the pond (that triggers sediment removal) is 3 ft. The number of years before the sediment must be removed from the pond is most nearly:

A. 1.8
B. 3.5
C. 4.3
D. 4.8

416

A watershed (area = 370 acres) is subdivided into five distinct land use classifications, as shown in the following table. Storm runoff data have been abstracted into a set of intensity-duration-frequency curves. Using the NRCS method, the net runoff (in) from a 20-year storm with gross rainfall of 5.6 in is most nearly:

A. 3.6
B. 3.1
C. 1.5
D. 2.2

Region	Area (acres)	Land Use	Soil Type	Time for Overland Flow (min)	Curve Number	Rational Runoff Coefficient
A	80	Lawns: fair condition	B	30	69	0.4
B	80	Forest	C	45	45	0.2
C	50	Paved	B	15	98	0.9
D	90	Residential: 4 lots/acre	D	25	87	0.6
E	70	Forest	A	45	35	0.2

(*Continued on next page*)

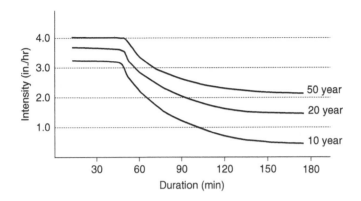

417

The following table shows recorded discharge at a stream monitoring station following a 2-hour storm. The tributary watershed area contributing runoff to the stream has been established as 115 acres.

Time (hr)	0	1	2	3	4	5	6
Discharge Q (ft³/sec)	23	84	127	112	75	32	25

The peak stream discharge (ft³/sec) that would be recorded following a 2-hour storm producing 1.7 in of runoff is most nearly:

A. 90

B. 65

C. 105

D. 127

418

A 230-acre watershed can be subdivided into five major parts based on land use and land cover.

Region	Area (acres)	Soil Type	Land Use	Overland Flow Time (min)
1	80	C	Single family homes on ½ acre lots	25
2	50	D	Lawns in good condition	42
3	10	B	Paved streets and sidewalks	15
4	50	C	Grassy areas: fair condition	34
5	40	A	Woods: fair condition	40

The composite NRCS curve number for this drainage basin is most nearly:
A. 55
B. 64
C. 73
D. 82

419

The parking lot shown below is located in Reno, Nevada. The concrete pavement slopes toward the 400-ft-long gutter in the center of the lot. Rain falling on the parking lot drains to the gutter. The gutter flow drains into a storm sewer inlet at one end. Only runoff from the parking lot enters this storm sewer inlet. Adjacent land drains elsewhere. For a significant rainfall event, the mean sheet flow velocity across the concrete parking lot is estimated to be 0.5 ft/sec. The estimated mean flow velocity in the gutter is 2.0 ft/sec. The 50-year recurrence interval peak discharge inflow (ft^3/sec) to the storm sewer inlet is most nearly:
A. 13
B. 18
C. 23
D. 28

(Continued on next page)

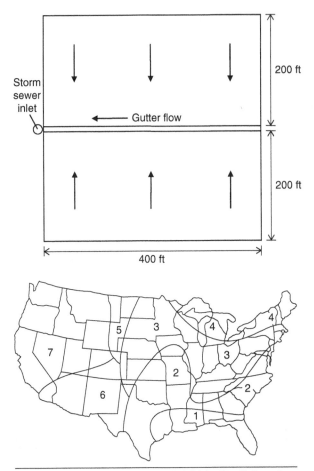

Geographical regions (continental U.S.) for Stool's formula.

Return Period (y)	Coefficients	Return						
		1	2	3	4	5	6	7
2	k	206	140	106	70	70	68	32
	b	30	21	17	13	16	14	11
5	k	247	190	131	97	81	75	48
	b	29	25	19	16	13	12	12
10	k	300	230	170	111	111	122	60
	b	26	29	23	16	17	23	13
25	k	327	260	230	170	130	155	67
	b	33	32	30	27	17	26	10
50	k	315	350	250	187	187	160	65
	b	28	38	27	24	25	21	8
100	k	367	375	290	220	240	210	77
	b	33	36	31	28	29	26	10

420

A stormwater detention pond is approximated as one acre in plan area with nearly vertical sides. The pond has a 5-ft-wide sharp-crested weir whose crest elevation is 125.0 ft above sea level. At a particular instant, the inflow to the pond was measured as 50 cfs and the surface elevation of the pond was 126.4 ft. Assuming the inflow rate stays constant, the outflow (cfs) from the pond 5 minutes later is most nearly:

A. 30
B. 49
C. 57
D. 65

421

A rectangular sharp-crested contracted weir is used to measure flow across a 20-ft-wide open channel. The weir opening is 6 ft wide. The weir head (upstream) is 5.4 ft. The channel discharge (ft^3/sec) is most nearly:

A. 835
B. 790
C. 250
D. 205

422

Which of the following statements is true?

    I. The effective stress in a moist soil immediately above the water table is calculated using the buoyant unit weight.

    II. The effective stress in a soil experiencing seepage flow can be either more or less than the effective stress with no seepage.

    III. The effective stress in a soil represents the pressure exerted by the pore water on the surface of the soil particles.

    IV. Increase in total stress in a soil is responsible for consolidation settlement.

A. II only

B. II and IV only

C. II, III, and IV only

D. I and III only

423

The average infiltration rate for buried water mains is given as 100 gpd per in diameter per mi. A water distribution network consists of the following pipe inventory:

Diameter (in)	Length (ft)
8	13,400
12	7,500
20	4,000
36	3,000

The extraneous flow rate (ft^3/sec) due to infiltration is most nearly:

A. 0.034

B. 0.023

C. 0.017

D. 0.011

424

A 30-ft-deep aquifer is confined between two impermeable layers of rock as shown in the following figure. The elevation of the piezometric surface is 20 ft above the top of the aquifer. A 9-in-diameter well is used to establish a steady-state pumping rate of 2,000 gal/min. The hydraulic conductivity of the soil in the aquifer is 1,000 ft/day. Observation wells 1 and 2 are drilled at radial distance 30 ft and 180 ft from the centerline of the pumping well as shown. The drawdown of the piezometric surface at observation well no. 1 is 4.5 ft. The drawdown of the piezometric surface (ft) at observation well no. 2 is most nearly:

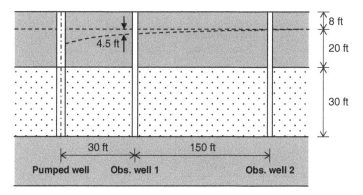

A. 2.1
B. 2.8
C. 3.2
D. 3.8

425

A daily pattern (typical) of wastewater flow rates influent into a wastewater treatment plant is summarized in the following table. The flow undergoes hydraulic stabilization by first passing through an equalization tank. The minimum volume (gallons) of the equalization tank is most nearly:

A. 2.6×10^5

B. 6.3×10^5

C. 1.5×10^6

D. 2.9×10^6

Time Period (hr)	Average Inflow Rate (ft³/sec)
00:00–02:00	18.3
02:00–04:00	12.4
04:00–06:00	8.7
06:00–08:00	7.3
08:00–10:00	6.4
10:00–12:00	8.9
12:00–14:00	14.5
14:00–16:00	18.9
16:00–18:00	6.5
18:00–20:00	5.6
20:00–22:00	8.3
22:00–24:00	13.2

426

A 700,000 gallon elevated water tank is used to provide firefighting water demand for a school complex. The total effective area of the school is 160,000 ft^2. The school buildings are constructed of masonry. The time of fire protection (hr) that the tank can provide is most nearly:

A. 1.0

B. 1.5

C. 2.0

D. 2.5

427

The schematic of an activated sludge process is shown in the following figure. The influent flow rate is 4 MGD. The primary clarifier removes 65% of the total suspended solids and 20% of the BOD$_5$. The primary sludge contains 6% solids. The quantity of primary sludge (gallons per day) is most nearly:

A. 7,000

B. 9,000

C. 11,000

D. 14,000

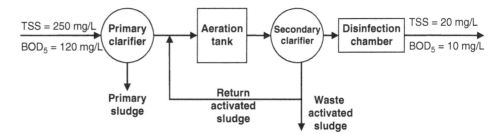

428

The flow net shown in the following figure describes the seepage into a long 20 ft deep and 40 ft wide excavation made in silty sand having a coefficient of permeability (K) equal to 3×10^{-4} cm/sec.

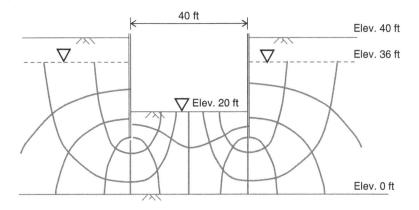

The total seepage flow (ft^3/min/ft) into the excavation trench is most nearly:

A. 1.0×10^{-3}

B. 3.0×10^{-3}

C. 8.0×10^{-3}

D. 15.0×10^{-3}

429

Methylene chloride or dichloromethane is a widely used solvent. It poses a hazard via inhalation and absorption through the skin. The reference dose (R_fD) for methylene chloride is 0.06 mg per kg body weight per day (mg/kg/d) based on liver toxicity in rats.

For an adult with a body weight of 70 kg with a lifetime water consumption of 2 L/day, the drinking water equivalent level (mg/L) is most nearly:

A. 0.06

B. 0.2

C. 0.6

D. 2.0

430

A municipal wastewater treatment plant treats a flow rate of 3 MGD containing $BOD_5 = 240$ mg/L. BOD removal takes place in a single-stage rock media trickling filter. Filter dimensions are: diameter = 80 ft, depth = 6 ft. Recirculation rate $R = 3$. The BOD_5 of the effluent is most nearly:

A. 60

B. 80

C. 120

D. 180

431

An industrial plant produces wastewater with the following characteristics:

Flow rate = 2 MGD
Temperature = 38°C
Ultimate BOD = 280 mg/L
Dissolved oxygen = 2.1 mg/L
Lead concentration = 0.5 mg/L

The stream into which the plant plans to discharge its wastewater has the following characteristics:

Flow rate = 30 ft³/sec
Temperature = 14°C
Ultimate BOD = 10 mg/L
Lead concentration = 4 µg/L

If the EPA limit for lead in surface waters is 15 µg/L, the level of pretreatment (%) necessary at the plant (before discharging into the stream) is most nearly:

A. 75

B. 80

C. 85

D. 0

432

A stream has the following characteristics:

Flow rate = 18 ft³/sec
Average velocity = 4 ft/sec
Temperature = 12°C
BOD_5 = 2.0 mg/L
Deoxygenation rate constant (log 10 at 20°C) = 0.20 day⁻¹
Reoxygenation rate constant (log 10 at 20°C) = 0.30 day⁻¹
D.O. = 5.1 mg/L

A factory discharges a wastewater stream into the river at point A. The wastewater has the following characteristics:

Flow rate = 750 gal/min
Temperature = 37°C
BOD_5 = 105 mg/L
D.O. = 1.7 mg/L

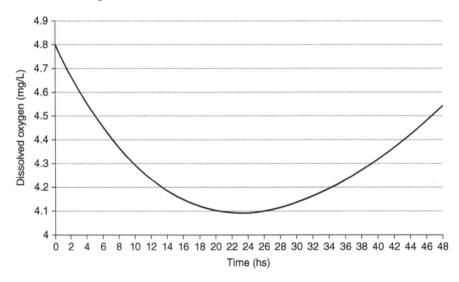

The dissolved oxygen sag curve for the stream-wastewater mix is shown in the curve shown here. The instant of mixing is considered $t = 0$. The oxygen deficit (mg/L) in the stream at a distance 20 mi downstream from point of mixing is most nearly:

A. 4.4
B. 6.4
C. 4.8
D. 4.1

433

The biochemical oxygen demand for a wastewater sample at 5 days is 4.5 mg/L and at 20 days is 8.3 mg/L. The deoxygenation rate constant (base 10, 20°C) is 0.1 day^{-1}. The nitrogenous oxygen demand (mg/L) of the sample is most nearly:

A. 3.8
B. 6.6
C. 0.0
D. 1.8

434

A water sample yields the following results:

Ca^{++}	60.0 mg/L
Mg^{++}	21.2 mg/L
Fe^{++}	2.2 mg/L
Na^+	5.5 mg/L
HCO_3^-	221.3 mg/L
Cl^-	33.5 mg/L
NO_3^-	21 mg/L
Turbidity	1.3 NTU
Odor	2.7 TON
Total coliform	1.7 MPN
TDS	437 mg/L

(Continued on next page)

Which of the following contaminants exceed the EPA's primary drinking water standards?

A. Nitrate and turbidity
B. Turbidity, total dissolved solids, and total coliform
C. Odor, total coliform, and TDS
D. Total coliform and turbidity

435

A 200 mL water sample is filtered through a standard Whatman filter. The filtrate is evaporated at 105°C in an evaporation dish (mass = 45.675 g). Following evaporation, the mass of the evaporation dish + solids = 47.225 g. The dish + solids are then ignited at 550°C, following which the mass of the evaporation dish + solids = 46.201 g.

The following data are also given for the solids retained on the filter paper.

Mass of crucible and filter paper = 25.334 g
Mass of crucible, filter paper, and dry (evaporated at 105°C) solids = 25.645 g
Mass of ignited crucible, filter paper, and solids = 25.501 g

The volatile dissolved solids concentration (mg/L) of the waste sample is most nearly:

A. 1,300
B. 1,800
C. 2,600
D. 5,100

436

A stream carries a flow of 5 MGD and has the following characteristics:

Temperature $= 12°C$
pH $= 7.8$
Dissolved oxygen concentration $= 5.7$ mg/L

A wastewater flow of 2,500 gpm having a temperature of 35°C and pH 3.4 is discharged into the stream. The pH immediately downstream of the mixing location is most nearly:

A. 6.0
B. 5.1
C. 3.8
D. 3.1

437

Dissolved oxygen concentration in a stream changes continuously as a result of deoxygenation (caused by an organic load carried by wastewater) and reoxygenation (caused by the interaction of the stream and the atmospheric air). An empirical model suggested by the USGS suggests:

$$k_r = \frac{3.3V}{H^{1.33}}$$

where k_r = reaeration coefficient (day^{-1})
V = average stream velocity (ft/sec)
H = stream depth (ft)

A natural stream has the following parameters:

Flow rate $= 2,514$ ft^3/sec
Area of flow section $= 1,323$ ft^2
Wetted perimeter $= 517$ ft
Average depth of flow $= 3.8$ ft

The reaeration coefficient (day^{-1}) of the stream is most nearly:

A. 0.5
B. 0.8
C. 1.1
D. 1.4

438

A secondary effluent at a wastewater treatment plant flows at a rate of 3.5 MGD. The effluent has the following characteristics: $BOD_5 = 20$ mg/L, TSS = 35 mg/L, pH = 8.5, temp = 35°C. The secondary effluent is disinfected by chlorination with a desired 2-log inactivation. Chlorine demand of the secondary effluent has been estimated as 3.5 mg/L. If the chlorine dose is 5.0 mg/L, the minimum required volume of the chlorination chamber (ft^3) is most nearly:

A. 6,100

B. 7,200

C. 8,300

D. 9,500

CT Values for Various Levels of Disinfection

Chlorine Residual Conc. (mg/L)	CT (mg/L-min) for Log-Inactivation				
	1.0	**1.5**	**2.0**	**2.5**	**3.0**
0.5	24	28	33	39	46
1.0	22	26	31	36	43
1.5	19	23	28	33	40
2.0	16	20	25	30	36
2.5	14	18	23	28	34
3.0	12	16	21	25	31
3.5	10	14	19	22	28

439

A water sample yields the following results:

Ca^{++}	60.0 mg/L
Mg^{++}	21.2 mg/L
Fe^{++}	2.2 mg/L
HCO_3^-	221.3 mg/L
Cl^-	33.5 mg/L

The hardness of the water sample, (mg/L as $CaCO_3$), is most nearly:

A. 240

B. 243

C. 510

D. 230

440

For a community, it is estimated that a population's water consumption will double over the next 20 years. The cost of expanding the existing water supply system will be compared to a phased program of expansion. Immediate development would cost $420,000 with annual maintenance costs of $40,000. A phased program would involve an initial investment of $200,000 and an estimated expenditure of $320,000 in 10 years. Annual maintenance cost under the phased program is estimated to be $20,000 for the first 10 years and $16,000 following that. Assume a *perpetual period of service* for each system and MARR = 7%. The ratio of the cost for phased program relative to the single investment program is most nearly:

A. 0.63

B. 0.95

C. 1.27

D. 1.62

END OF WATER RESOURCES & ENVIRONMENTAL DEPTH EXAM

USE THE ANSWER SHEET ON THE NEXT PAGE

Water Resources & Environmental Depth Exam: Answer Sheet

401	Ⓐ	Ⓑ	Ⓒ	Ⓓ
402	Ⓐ	Ⓑ	Ⓒ	Ⓓ
403	Ⓐ	Ⓑ	Ⓒ	Ⓓ
404	Ⓐ	Ⓑ	Ⓒ	Ⓓ
405	Ⓐ	Ⓑ	Ⓒ	Ⓓ
406	Ⓐ	Ⓑ	Ⓒ	Ⓓ
407	Ⓐ	Ⓑ	Ⓒ	Ⓓ
408	Ⓐ	Ⓑ	Ⓒ	Ⓓ
409	Ⓐ	Ⓑ	Ⓒ	Ⓓ
410	Ⓐ	Ⓑ	Ⓒ	Ⓓ
411	Ⓐ	Ⓑ	Ⓒ	Ⓓ
412	Ⓐ	Ⓑ	Ⓒ	Ⓓ
413	Ⓐ	Ⓑ	Ⓒ	Ⓓ
414	Ⓐ	Ⓑ	Ⓒ	Ⓓ
415	Ⓐ	Ⓑ	Ⓒ	Ⓓ
416	Ⓐ	Ⓑ	Ⓒ	Ⓓ
417	Ⓐ	Ⓑ	Ⓒ	Ⓓ
418	Ⓐ	Ⓑ	Ⓒ	Ⓓ
419	Ⓐ	Ⓑ	Ⓒ	Ⓓ
420	Ⓐ	Ⓑ	Ⓒ	Ⓓ

421	Ⓐ	Ⓑ	Ⓒ	Ⓓ
422	Ⓐ	Ⓑ	Ⓒ	Ⓓ
423	Ⓐ	Ⓑ	Ⓒ	Ⓓ
424	Ⓐ	Ⓑ	Ⓒ	Ⓓ
425	Ⓐ	Ⓑ	Ⓒ	Ⓓ
426	Ⓐ	Ⓑ	Ⓒ	Ⓓ
427	Ⓐ	Ⓑ	Ⓒ	Ⓓ
428	Ⓐ	Ⓑ	Ⓒ	Ⓓ
429	Ⓐ	Ⓑ	Ⓒ	Ⓓ
430	Ⓐ	Ⓑ	Ⓒ	Ⓓ
431	Ⓐ	Ⓑ	Ⓒ	Ⓓ
432	Ⓐ	Ⓑ	Ⓒ	Ⓓ
433	Ⓐ	Ⓑ	Ⓒ	Ⓓ
434	Ⓐ	Ⓑ	Ⓒ	Ⓓ
435	Ⓐ	Ⓑ	Ⓒ	Ⓓ
436	Ⓐ	Ⓑ	Ⓒ	Ⓓ
437	Ⓐ	Ⓑ	Ⓒ	Ⓓ
438	Ⓐ	Ⓑ	Ⓒ	Ⓓ
439	Ⓐ	Ⓑ	Ⓒ	Ⓓ
440	Ⓐ	Ⓑ	Ⓒ	Ⓓ

6

Transportation Depth Exam

The following set of questions numbered 501 to 540 is representative of a 4-hr transportation depth exam according to the syllabus and guidelines for the Principles and Practice (P&P) of Civil Engineering Examination administered by the National Council of Examiners for Engineering and Surveying (NCEES), current for the October 2020 examination.

Enter your answers on page 149. Detailed solutions are on page 279.

501

A study area is divided into three socio-economic zones, whose trip production and trip attractions are tabulated as follows:

Zone	1	2	3	Total
Trip Productions	250	450	440	1,140
Trip Attractions	120	350	670	1,140

Resistance to travel between zones is represented by the following matrix of friction factors, which are approximately inversely proportional to the time of travel between zones.

Zone	1	2	3
1	40	90	75
2	90	25	35
3	75	35	40

The number of trips produced by zone 3 and attracted to zone 2, according to the first iteration of the gravity model, is most nearly:

A. 110

B. 130

C. 150

D. 170

502

A single-lane roundabout (single entry lane conflicting with a single circulating lane) has a conflicting volume = 230 pc/h. The capacity (pc/h) of the entering lane, adjusted for heavy vehicles is most nearly:

A. 865

B. 900

C. 945

D. 1010

503

A six-lane freeway through rolling terrain (rural) has 12-ft lanes, a full cloverleaf interchange every 1.25 mi on average and 8-ft-wide shoulders. Traffic studies have resulted in the following data:

ADT = 65,000 veh/day.
Terrain is level.
The traffic stream includes 8% trucks, 3% buses, and 5% RVs.
K = 0.12
Peak hour factor, based on 15-min traffic counts, is 0.9.
During peak flow, directional split = 60/40.

The LOS for the peak direction of the freeway is most nearly:

A. A
B. B
C. C
D. D

504

A six-lane highway has the following characteristics:

Level terrain
Lane width = 11-ft lanes
Posted speed limit = 50 mph
Average spacing between accessing driveways = 600 ft
Right shoulder width = 6 ft.
Clear distance to nearest obstruction in median = 4 ft
Directional hourly volume = 3,440 veh/hr
The traffic stream includes 8% trucks, 3% buses, and 2% RVs
Peak hour factor, based on 15-min traffic counts, is 0.88
Drivers are mostly commuters.

The LOS for the peak direction of the highway is most nearly:

A. E
B. D
C. C
D. B

505

A two-lane Class I highway in rolling terrain has the following characteristics:

Lane width = 12 ft, shoulder width = 6 ft
40% no passing zones
Two-directional flow rate = 2800 pc/h
Directional split = 70/30
10% trucks and buses, 6% RVs
PHF = 0.85

For the design direction of the highway, the average travel speed has been calculated as 42 mph and the percent time spent following has been calculated as 63%.

The level of service for the design direction is:

A. B
B. C
C. D
D. F

506

A parking lot is open during the hours of 8AM–6PM. During this time, exactly 360 cars were parked on the lot—10% for 1 hour, 15% for 2 hours, 20% for 3 hours, 30% for 4 hours, and the remaining for the entire day. On average, 15% of the spaces are vacant and the operational efficiency factor is 80%.

The space-hour demand and number of parking spaces in the lot are most nearly:

A. 1,990 space-hours and 250 spaces
B. 1,690 space-hours and 240 spaces
C. 1,690 space-hours and 170 spaces
D. 1,990 space-hours and 200 spaces

507

A traffic stream has mean headways of 2.4 seconds. If the jam density is 64 veh/mile, the optimum speed is 50 mph and the flow is 1500 veh/hr, the capacity of the highway (veh/hr) according to the Greenshield model is most nearly:

A. 1,530
B. 1,600
C. 1,640
D. 1,720

508

A four-leg intersection in an urban location had 15 recorded accidents (all types) during 2005. During this year, the average daily volumes entering the intersection from the four approaches were 1,900, 1,270, 1,620, and 930 vehicles. The intersection's accident rate (per hundred million entering vehicles) is most nearly:

A. 7.2
B. 72
C. 720
D. 7,200

509

A stretch of highway experienced 12 crashes during 2008. ADT during 2008 was 15,600. Due to the significant number of crashes, the following three strategies were put in place:

1. Lane widening—expected CMF = 0.90
2. Parking restrictions—expected CMF = 0.75
3. Lighting improvement—expected CMF = 0.80

If traffic grows by approximately 3% every year, the number of crashes expected in 2010 is most nearly:

A. 5
B. 7
C. 9
D. 11

510

A car accelerates uniformly from rest to its peak speed of 70 mph. The acceleration rate is 8 mph/sec. After cruising a certain distance at peak speed, the vehicle brakes to rest, decelerating uniformly at 10 mph/sec. If the total distance traveled is 0.5 mi, the average running speed (mph) is most nearly:

A. 48.7
B. 51.2
C. 53.6
D. 57.4

511

Pedestrian movements along a sidewalk (shown shaded gray in the following figure) city block can be divided into four zones demarcated by points A–E and described as follows:

Zone	Description	Length (ft)	Pedestrian Space (ft²/p)
AB	Intersection	25	10
BC	Residential	150	50
CD	Commercial	210	30
DE	Intersection	25	12

The pedestrian space (ft²/p) for the entire block is most nearly:

A. 43
B. 40
C. 35
D. 28

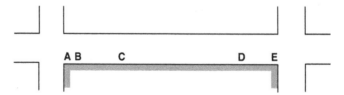

512

A circular horizontal curve has PC at 12 + 43.56. Degree of curve = 4°. Deflection angle between tangents is 56°24′45″ (right). In order to meet lateral clearance requirements, it is proposed to shift the back tangent by a parallel offset of 20 ft, while maintaining the PC, as shown in the following figure. The new degree of curve is most nearly:

A. 3.7
B. 3.8
C. 3.9
D. 4.1

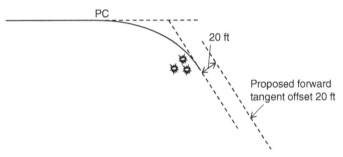

513

A horizontal compound curve connects two tangents as shown. The bearing of the back tangent is N 46°25′32″E and the bearing of the forward tangent is S 17°56′21″E. The curve radii to be used are $R_1 = 1,800'$ and $R_2 = 900'$. The PC is located at station 138 + 34.12 and the PI at station 158 + 12.98.

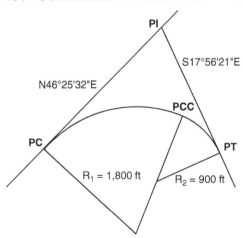

(*Continued on next page*)

The station of the PCC is most nearly:

A. 145 + 12.54

B. 147 + 05.23

C. 148 + 50.61

D. 151 + 03.32

514

A horizontal circular curve is to connect a back-tangent bearing S42°30′W to a forward tangent bearing N70°W. If the degree of curve is 3°45′ and the tangents intersect at station 50 + 22.30, the deflection angle for station 57 + 00.00 is most nearly:

A. 31°50′56″

B. 21°40′28″

C. 41°32′15″

D. 27°40′24″

515

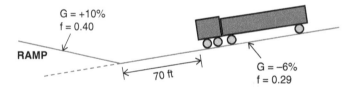

A runaway truck (gross vehicle weight 90,000 lb) is traveling at 65 mph down a 6% grade when the driver sees an arrester ramp 70 ft ahead. If the brakes are applied while 70 ft upstream of the ramp and braking is maintained while the exit ramp is taken, the distance traveled on the ramp (ft) before the truck comes to a stop is most nearly:

A. 150

B. 190

C. 250

D. 320

516

A parabolic crest curve is followed by a sag curve. The two curves are connected by a tangent section as shown. A bridge structure is located at station 40 + 55.00. The elevation of the low point on the bridge is 405.54 ft. The design speed (mph) of a truck (assume driver's eye elevation = 8 ft) based on sight distance on curve number 2 is most nearly:

A. 60
B. 70
C. 80
D. 90

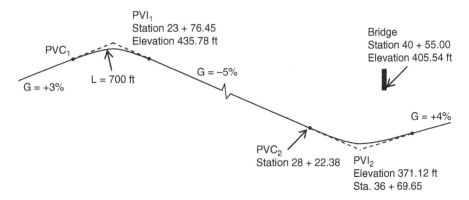

517

A parabolic vertical curve joins a grade of −5% to a grade of +3%. The PVC is at station 53 + 12.50 and the PVI is at station 60 + 09.00. Elevation of the PVI is 365.57 ft. The curve passes under a bridge structure at station 55 + 05.20. The bottom elevation of the bridge is 405.20 ft. The vertical clearance under the bridge (ft) is most nearly:

A. 12.5
B. 13.4
C. 14.2
D. 14.9

518

A parabolic vertical curve is being designed to connect two grades with $G_1 = +2.1\%$ and $G_2 = -1.5\%$. If the design speed is 65 mph. the minimum length of vertical curve (ft) to satisfy AASHTO safe stopping distance criteria is most nearly:

A. 600
B. 650
C. 800
D. 700

519

The vertical profile of a bridge constructed over a creek follows a parabolic vertical curve of length 950 ft as shown in the following figure. The PVC of the curve is at elevation 247.65 ft above sea level and the 100-year flood elevation is at 238.23 ft A.S.L. The curve connects a grade of −3% to a grade of +5%. The minimum vertical clearance (ft) above the 100-year flood is most nearly:

A. 3.25
B. 3.50
C. 3.75
D. 4.00

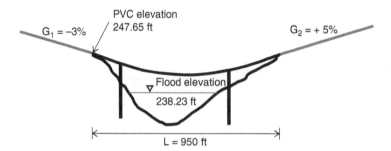

520

A three-legged intersection is shown in the following figure. The minor street approach to the intersection has two lanes, is controlled by a STOP sign, has a grade of +4% heading into the intersection, and serves 15% of trucks during the peak hour. The major street has a basically flat grade, a speed limit of 45 mph, and serves 20% of trucks during the peak hour. The median is 6 ft wide.

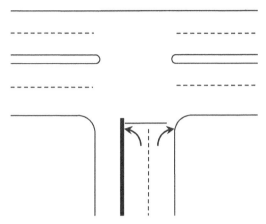

The critical gap (seconds) for the left turn movement from the minor street during the peak hour is most nearly:

A. 11 sec

B. 10 sec

C. 9 sec

D. 8 sec

521

A two-lane freeway segment with a single-lane on ramp followed by a single-lane off ramp is shown in the following figure. All indicated volumes are hourly volumes. PHF on the freeway is 0.88 and on the ramps is 0.95.

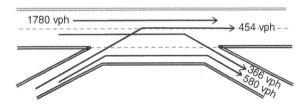

1780 vph

454 vph

366 vph

580 vph

The volume ratio for the weaving segment is most nearly:

A. 0.25

B. 0.35

C. 0.65

D. 0.75

522

A three-lane freeway interchange in a suburban area has a one-lane diverging ramp. The adjusted peak hour volume just downstream from the diverge of interest is 4,100 passenger cars, while the adjusted peak hour volume on the ramp is 500 passenger cars. The deceleration lane is 1,100 ft long, the ramp free-flow speed is 54 mph, and the ramp has a +3% grade. An off ramp 1,250 ft downstream of the diverge of interest serves an adjusted peak hour volume of 450 passenger cars.

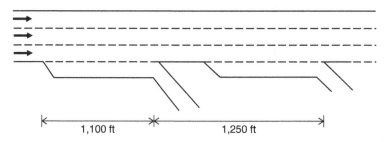

The adjusted traffic volume in the right two lanes of the freeway just upstream of the diverge of interest is most nearly:

A. 2,700
B. 2,850
C. 3,050
D. 3,250

523

A two-lane highway has a posted speed limit of 50 mph. The alignment consists of a dangerous curve where the maximum posted speed is 20 mph. What is the distance (ft) upstream of the low-speed limit section where a warning sign must be posted?

A. 125
B. 150
C. 175
D. 200

524

A two-lane rural freeway segment has 12-ft lanes and an 8-ft shoulder. A cut slope of 1V:4H exists adjacent to the shoulder as shown in the figure. The ADT is 16,000 and the design speed is 65 mph. The minimum width (ft) measured from the edge of the shoulder, which should be free of any obstructions is most nearly:

A. 18
B. 26
C. 38
D. 46

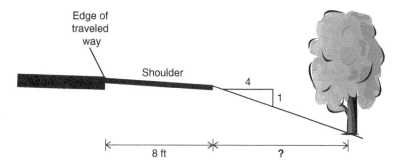

525

Which of the following is NOT an objective for traffic barriers?
A. Increasing capacity
B. Separating opposing flow
C. Channeling various modes of traffic flow
D. Work zone safety

526

A one-lane major street intersects with a one-lane minor street at an at-grade intersection. Traffic counts were collected during 12 consecutive hours of a typical day.

	Major Street		Minor Street	
Time	NB	SB	EB	WB
6 AM–7 AM	410	135	45	55
7 AM–8 AM	725	320	110	160
8 AM–9 AM	650	430	100	180
9 AM–10 AM	570	500	80	100
10 AM–11 AM	450	530	90	110
11 AM–12 PM	450	550	80	110
12 PM–1 PM	550	600	120	110
1 PM–2 PM	410	620	160	180
2 PM–3 PM	390	640	130	150
3 PM–4 PM	350	710	160	190
4 PM–5 PM	400	750	200	190
5 PM–6 PM	420	720	170	150

Which of the following statements are true?

A. Neither warrant 1 nor warrant 2 are met.

B. Both warrants 1 and 2 are met.

C. Warrant 1 is met but warrant 2 is not.

D. Warrant 1 is not met but warrant 2 is met.

527

According to ADA accessibility guidelines, the minimum number of handicapped-accessible spaces in a parking lot which accommodates a total of 630 vehicles is most nearly:

A. 7
B. 9
C. 10
D. 13

528

What is the recommended length of cycle for a four-phase signal with 3 seconds lost time per phase, with the following critical movements?

Phase	Lane Group Volume (vph)	Saturation Volume for Lane Group (vphg)	Green Time Fraction
1	100	900	0.11
2	600	1,900	0.36
3	105	900	0.13
4	630	2,300	0.40

A. 105 sec
B. 125 sec
C. 140 sec
D. 160 sec

The following diagram and the associated data are to be used for problems 529–531.

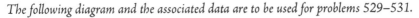

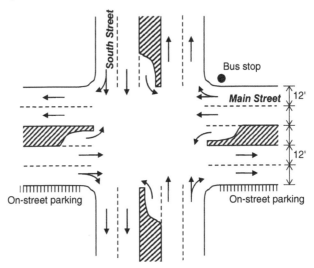

A right-angle intersection between two four-lane highways (Main St and South St) is shown. The 80-sec signal cycle has four phases: (1) EBL and WBL, (2) EBTH/R and WBTH/R, (3) NBL and SBL, and (4) NBTH/R and SBTH/R. Each yellow time separating phases is 3 sec and each all-read interval between phases is 1 sec. The critical v/s ratios for each phase are listed as follows:

Phase	Critical Movement	v/s
1	WBL	0.232
2	EBTH/R	0.156
3	NBL	0.150
4	SBTH/R	0.135

Posted speed on both approaches = 40 mph
Heavy vehicles = 5%
PHF = 0.92
25% of TH/R traffic is RT.
10 buses per hour on EBTH/R and WBTH/R approaches
25 parking maneuvers per hour for EBTH/R and WBTH/R approaches
Perception-reaction time = 1 sec
Deceleration rate = 10 ft/s^2
Vehicle length = 20 ft

Pedestrian volumes: on north-south walkways (10 ft wide) = 1,200 pedestrians/hr
on east-west walkways (10 ft wide) = 800 pedestrians/hr

529

The saturation flow (veh/hr) for the WBTH/R lane group is most nearly:

A. 2,650
B. 2,860
C. 3,030
D. 3,210

530 (See figure and data associated with problem 529)

The minimum clearance interval (yellow + all red) (sec) for the intersection is most nearly:

A. 4.0
B. 4.5
C. 5.3
D. 5.8

531 (See figure and data associated with problem 529)

The minimum green time for the north-south signal phase, based on pedestrian volume, is most nearly:

A. 25 sec
B. 22 sec
C. 19 sec
D. 16 sec

532

An agency is investigating installing a signal at an intersection in a large city where the 85th percentile speeds are 40 mph on both approaches. The intersection has two lanes in the major direction and one lane in the minor direction. Recent traffic counts from the high-volume hours at the intersection are provided as follows:

Time Period	Vehicles on Both Major Approaches Combined	Vehicles on Highest Volume Minor Street Approach
7:00–8:00 AM	600	210
8:00–9:00 AM	650	230
11:00–12:00 AM	550	210
12:00–1:00 PM	730	250
3:00–4:00 PM	660	260
4:00–5:00 PM	830	275
5:00–6:00 PM	990	310
6:00–7:00 PM	800	250

Which warrant(s) from the 2009 *MUTCD* does this intersection meet?

A. Warrants 1 and 2
B. Warrant 2 only
C. Warrants 1 and 3
D. Warrants 2 and 3

533

According to the Manual of Uniform Traffic Control Devices, 2009, a temporary traffic control designated "long-term stationary" is one that occupies a fixed work zone for a period exceeding:

A. 3 days
B. 4 days
C. 5 days
D. 7 days

534

The following data is given for a soil sample:

Sieve Analysis:		Atterberg Tests:

Sieve size	Percent Retained
No. 4	8
No. 10	10
No. 20	12
No. 40	21
No. 100	15
No. 200	8

Atterberg Tests:
Liquid Limit = 43
Plastic Limit = 21

The AASHTO soil classification is:
A. A-6
B. A-7
C. A-2-6
D. A-2-7

535

Results from a standard Proctor compaction test from six soil samples from a borrow pit are tabulated in the following table. The natural moisture content of the excavated material is 12%. The fill location requires 1.5 million yd^3 of soil compacted to a minimum 90% of the maximum Proctor dry density.

Sample	Net Weight of Soil (lb)	Moisture Content (%)
1	3.24	12
2	3.70	14
3	3.95	16
4	4.21	18
5	3.90	20
6	3.40	22

(*Continued on next page*)

The total volume of borrow soil that must be excavated (yd³) is most nearly:

A. 1.72 million

B. 1.65 million

C. 1.53 million

D. 1.42 million

536

A four-lane divided highway (two lanes in each direction) currently has ADT of 23,000 vehicles (each way). Approximately 12% of the traffic consists of trucks. Traffic is expected to grow by a 4% annual rate over the next 20 years. Table 1 shows data collected from a weigh station during a typical day for 1,078 trucks.

Table 1: Traffic Data (Axle Counts)		
Axle Load (k)	Number of Single Axles	Number of Tandem Axles
10	2,420	
14	630	
18	301	
20	22	
22	6	24
25	1	15
28		12
32		11

Table 2: Load Equivalence Factors	
Single Axles	
Axle Load (k)	Load Equivalence Factor
10	0.0877
14	0.360
18	1.0
20	1.51
22	2.18
25	3.53
Tandem Axles	
Axle load (k)	Load Equivalence Factor
22	0.180
25	0.308
28	0.495
32	0.857

Using the Asphalt Institute Load Equivalence Factors (LEF) shown in Table 2, the design 20-year 18-k ESAL is most nearly:

A. 9×10^6

B. 15×10^6

C. 20×10^6

D. 23×10^6

537

Which of the following statement/s is/are true for a continuously reinforced-concrete pavement?

    I. Increasing pavement thickness typically decreases the density of punchouts.

    II. Punchouts are caused by compressive stresses in the slab due to wheel loads.

    III. Pavement durability is affected by the type of base used to support the pavement slab.

    IV. Durability of a pavement is directly related to a number of load cycles.

    V. The MEPDG does not allow the designer to design a pavement with a specific upper limit on crack density.

A. All of them

B. I, III, and IV only

C. I, III, IV, and V only

D. I and IV only

538

The following table shows recorded discharge at a stream monitoring station following a 2-hr storm. The tributary watershed area contributing runoff to the stream has been established as 115 acres.

Time (hr)	0	1	2	3	4	5	6
Discharge Q (ft³/sec)	23	84	127	112	75	32	25

The peak stream discharge (ft³/sec) that would be recorded following a 2-hr storm producing 1.7 in of runoff is most nearly:

A. 90

B. 80

C. 180

D. 160

539

A 36-in diameter reinforced-concrete pipe under a roadway serves as a culvert operating under inlet control. The following data apply to the culvert:

Upstream invert: Elevation 325.0
Downstream invert: Elevation 322.5
Length of culvert: 145 ft
Entrance condition: Headwall with square edge
Roadway elevation: 337.8 ft
Culvert flow rate due to 100-year flood event: 100 cfs

The minimum vertical clearance (ft) between the roadway and the 100-year flood elevation is most nearly:

A. 5.2
B. 2.2
C. 4.8
D. 7.8

540

In order to meet current demand, a 1-mi segment of an urban highway must be widened by two lanes. The cost of the improvement will be compared to a phased program of expansion. Immediate development would cost $4,200,000 with annual maintenance costs of $40,000. A phased program would involve an initial investment of $2,000,000 and an estimated expenditure of $3,200,000 in 10 years. Annual maintenance cost under the phased program is estimated to be $28,000 for the first 10 years and $36,000 following that. Assume a *perpetual period of service* for each system and MARR of 7%. The ratio of the cost for phased program relative to the single investment program is most nearly:

A. 0.85
B. 0.95
C. 1.04
D. 1.17

END OF TRANSPORTATION DEPTH EXAM

USE THE ANSWER SHEET ON THE NEXT PAGE

TRANSPORTATION PM Exam: Answer Sheet

501	Ⓐ	Ⓑ	Ⓒ	Ⓓ
502	Ⓐ	Ⓑ	Ⓒ	Ⓓ
503	Ⓐ	Ⓑ	Ⓒ	Ⓓ
504	Ⓐ	Ⓑ	Ⓒ	Ⓓ
505	Ⓐ	Ⓑ	Ⓒ	Ⓓ
506	Ⓐ	Ⓑ	Ⓒ	Ⓓ
507	Ⓐ	Ⓑ	Ⓒ	Ⓓ
508	Ⓐ	Ⓑ	Ⓒ	Ⓓ
509	Ⓐ	Ⓑ	Ⓒ	Ⓓ
510	Ⓐ	Ⓑ	Ⓒ	Ⓓ
511	Ⓐ	Ⓑ	Ⓒ	Ⓓ
512	Ⓐ	Ⓑ	Ⓒ	Ⓓ
513	Ⓐ	Ⓑ	Ⓒ	Ⓓ
514	Ⓐ	Ⓑ	Ⓒ	Ⓓ
515	Ⓐ	Ⓑ	Ⓒ	Ⓓ
516	Ⓐ	Ⓑ	Ⓒ	Ⓓ
517	Ⓐ	Ⓑ	Ⓒ	Ⓓ
518	Ⓐ	Ⓑ	Ⓒ	Ⓓ
519	Ⓐ	Ⓑ	Ⓒ	Ⓓ
520	Ⓐ	Ⓑ	Ⓒ	Ⓓ

521	Ⓐ	Ⓑ	Ⓒ	Ⓓ
522	Ⓐ	Ⓑ	Ⓒ	Ⓓ
523	Ⓐ	Ⓑ	Ⓒ	Ⓓ
524	Ⓐ	Ⓑ	Ⓒ	Ⓓ
525	Ⓐ	Ⓑ	Ⓒ	Ⓓ
526	Ⓐ	Ⓑ	Ⓒ	Ⓓ
527	Ⓐ	Ⓑ	Ⓒ	Ⓓ
528	Ⓐ	Ⓑ	Ⓒ	Ⓓ
529	Ⓐ	Ⓑ	Ⓒ	Ⓓ
530	Ⓐ	Ⓑ	Ⓒ	Ⓓ
531	Ⓐ	Ⓑ	Ⓒ	Ⓓ
532	Ⓐ	Ⓑ	Ⓒ	Ⓓ
533	Ⓐ	Ⓑ	Ⓒ	Ⓓ
534	Ⓐ	Ⓑ	Ⓒ	Ⓓ
535	Ⓐ	Ⓑ	Ⓒ	Ⓓ
536	Ⓐ	Ⓑ	Ⓒ	Ⓓ
537	Ⓐ	Ⓑ	Ⓒ	Ⓓ
538	Ⓐ	Ⓑ	Ⓒ	Ⓓ
539	Ⓐ	Ⓑ	Ⓒ	Ⓓ
540	Ⓐ	Ⓑ	Ⓒ	Ⓓ

7

Construction Depth Exam

The following set of questions numbered 601 to 640 is representative of a 4-hr construction depth exam according to the syllabus and guidelines for the Principles and Practice (P&P) of Civil Engineering Examination administered by the National Council of Examiners for Engineering and Surveying (NCEES), current for the October 2020 examination.

Enter your answers on page 175. Detailed solutions are on page 297.

601

The following table shows areas of cut and fill sections at locations 50 ft apart. Assume shrinkage = 12% and bulking = 25%. Cumulative earthwork (yd³) between stations 0 + 0.00 and 3 + 0.00 is most nearly:

A. 1,340
B. 1,850
C. 2,200
D. 2,500

Station	Area (ft²)	
	Cut	Fill
0 + 00.00	563.2	342.2
0 + 50.00	213.5	213.6
1 + 00.00	123.5	343.3
1 + 50.00	654.6	111.0
2 + 00.00	973.1	762.4
2 + 50.00	567.3	342.9
3 + 00.00	451.6	190.4

602

An excavation plan is outlined in the following figure. Each grid square is 50 ft × 50 ft. The numbers are depth of cut (ft) at indicated locations. The total volume of cut (yd³) is most nearly:

A. 4,500
B. 5,000
C. 5,500
D. 6,000

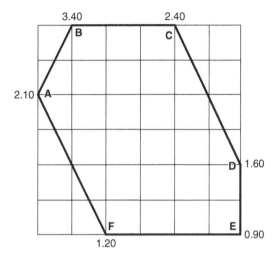

603

While conducting a site survey, a benchmark elevation of 154.45 ft above sea level is established. A level and rod arrangement measures the following:

Rod reading at benchmark = 7.85 ft
Rod reading at station A = 8.92 ft

The elevation of station A (ft) is most nearly:

A. 171.22
B. 155.52
C. 153.38
D. 137.68

604

The following table shows ordinates of a mass haul diagram for a highway construction project. The free haul distance is 300 ft. Overhaul cost is $3.70/station yard.

Station	30 + 00	42 + 40	45 + 10	47 + 00	48 + 10	50 + 30	53 + 20
Cumulative Yardage (yd³)	0	800	1,275	1,580	1,275	680	0

The total cost of overhaul is most nearly:

A. $35,000

B. $40,000

C. $45,000

D. $50,000

605

A contractor has the following options for a project lasting 18 months:

Option A: Monthly rental of excavation equipment at $15,000 per month + operating costs $2,000/month

Option B: Purchase equipment for $200,000
Maintenance costs $8,000/month
Resale value of equipment after 18 months = $120,000

Nominal interest rate = 10% p.a.

The benefit:cost ratio of option B (purchasing) is most nearly:

A. 1.95

B. 1.55

C. 1.25

D. 0.95

606

A structure has been examined to determine a need for rehabilitation. The related costs are summarized below:

Current annual costs = $40,000
Estimated rehabilitation cost = $350,000
Annual costs projected after rehabilitation = $15,000
Expected useful life remaining = 20 years
Projected increase in residual value (at end of useful life) = $200,000

The return on investment (ROI) for performing the rehabilitation is most nearly:

A. 5%
B. 6%
C. 7%
D. 8%

607

A 450-ft-long canal (trapezoidal section: 19-ft bottom width, 3H:2V side slopes) is to be lined with concrete to a nominal thickness of 8 in. Material waste is estimated to be 5% (by weight).

Concrete cost = $98/yd³
Concrete pour rate = 8 yd³/hr
Surface finishing compound is purchased in 5-gal containers costing $40 each
Coverage of surfacing compound = 300 ft²/gal

The total material cost ($) is most nearly:

A. 75,000
B. 85,000
C. 95,000
D. 105,000

608

An existing landfill is rectangular at the base with plan dimensions 4,000 ft × 2,500 ft. The sides slope 3H:1V. The height of the landfill is 12 ft. The top of the landfill must be covered with cover soil to a depth of 18 in. The amount of cover soil (yd³) required is most nearly:

A. 525,000

B. 650,500

C. 712,200

D. 821,435

609

The architect for a warehouse building has proposed a design change to ensure compatibility with local building codes. The revised SOW is to substitute one layer of ⅝-in-thick sheet rock with two layers of ½-in-thick fire code (FC) GWB and to provide insulation within the wall cavity.

The building plan dimensions are: 180 ft × 200 ft
Floor height: 12 ft
Openings: Eight 6 ft 0 in wide × 10 ft 0 in high openings.

Labor rates

Carpenter foreman (working)	$50.00 fully burdened
Carpenter (journeyman)	$40.00 fully burdened
Laborer	$25.00 fully burdened

Work crew

4 carpenters
2 laborers
1 Working foreman

Work crew productivity (based on 8 hr/day)

GWB installation	960 ft²/L.H.
Insulation	1,920 ft²/L.H.

Material costs (add 10% waste factor for all materials)

4 ft 0 in × 10 ft 0 in × ½ in GWB (FC)	$0.285/ft²
4 ft 0 in × 10 ft 0 in × ⅝ in GWB	$0.255/ft²
Insulation	$0.45/ft²
Contractor's overhead	10%
Contractor's profit	5%

The complete cost ($) of the change order for the revised SOW is most nearly:

A. 3,300

B. 4,200

C. 9,800

D. 7,800

610

A contract has 90 days remaining until completion. Early completion will be paid $18,000 bonus per day by which the duration is shortened. Activities on the critical path are targeted for "crashing." Critical activities can be accelerated by allocating additional crew units. The number of schedule days saved is not directly proportional to the number of crews added. For each crew unit added, the cost is $3,200 per day for labor and equipment. If crews must be added as a complete unit, the net bonus is maximized by the addition of how many crews?

A. 1 additional crew unit

B. 2 additional crew units

C. 3 additional crew units

D. None of the crashing schemes are profitable.

Crew Added	Added Crew Cost per Day	Schedule Days Reduced
1	$3,200	15
2	$6,000	21
3	$9,000	31

611

Dump trucks with a capacity of 16 yd^3 are used to dispose of excavated materials from a site. The distance from the dump site is 4 mi and the average speed of the dump trucks is 30 mph. The bucket capacity of the power shovel is 3 yd^3; power shovel excavation rate = 10 yd^3/min (bank measure) and the transfer time to the dump trucks (per 3 yd^3 load) is 40 sec and a dump time (per 3 yd^3 load) is 30 sec. The excavated material has a swell of 10%.

The factors affecting the job site productivity of the dump truck are: 0.80 for equipment idle time and 0.70 for operator efficiency. If a fleet of dump trucks are used to haul the excavated material, the number of trucks needed is most nearly:

A. 3

B. 4

C. 5

D. 6

612

The project manager for an excavation project is considering the costs of four different excavator models. The results are listed in the following table. The amount of material to be excavated is 18,000 yd^3.

Type	Production (yd³/hr)	Fixed Cost ($)	Variable Cost ($/hr)
1	320	12,000	55
2	280	10,000	50
3	360	17,000	55
4	300	12,000	45

The excavator which is most economical is:

A. Type 1

B. Type 2

C. Type 3

D. Type 4

613

A crate weighing 900 lb is being lifted by four cables attached to the corners as shown. The attachment point for the cables (O) is 8 ft directly above the center of gravity of the load. Tension in each cable (lb) is most nearly:

A. 270
B. 225
C. 342
D. 412

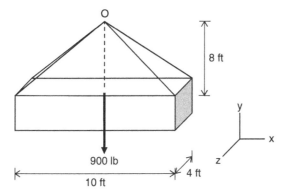

614

It is proposed to drive a concrete pile using a pile hammer with energy of 50,000 ft lb-ft. The minimum factor of safety against bearing capacity failure is equal to 6.0. The superstructure load transmitted to the pile is expected to be 40 tons.

The ENR pile driving equation gives the static (ultimate) bearing capacity of a pile based on a pile driving test as:

$$Q_{ult} = \frac{WH}{S + 1.0}$$

where Q_{ult} = ultimate capacity (lb)
$\quad W$ = weight of the ram (lb)
$\quad H$ = height the ram falls (in)
$\quad S$ = pile set (in per blow)

(*Continued on next page*)

The required blow count (blows per ft) to achieve a design capacity of 40 tons is most nearly:

A. 8 blows per ft
B. 28 blows per ft
C. 35 blows per ft
D. 50 blows per ft

615

A building construction site is represented by the rectangle ABCD shown in the following figure. The original elevation of the water table at the site is 325.8 ft (above sea level). The dashed line shows the limits of the proposed construction. The bottom elevation of the mat foundation for the building is 325.65 ft. Steady-state pumping from 9-in-diameter well at point B is to be used to lower the GWT. The transmissivity of the underlying aquifer is 250 ft²/hr. During construction, the GWT must be lowered to at least 5 ft below the bottom of the foundation and no higher than an elevation of 324.0 ft within the limits of the site. The minimum steady-state pumping rate (gal/min) required is most nearly:

A. 1,000
B. 1,500
C. 2,000
D. 3,000

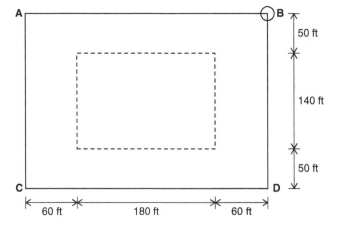

616

An excavator has a bucket capacity of 3.0 yd³. Its operation cycle consists of the following phases—(a) excavation time = 45 sec, (b) travel time (two-way) = 4 min, and (c) transfer time (to trucks) = 45 sec. The excavator transfers the excavated material to a fleet of trucks that carry the material offsite. The following data is given:

Overall efficiency factor for the excavator = 90%
Truck capacity = 15 yd³
Truck cycle time (transfer + two-way travel + dumping) = 90 min
The quantity of excavated material = 2,600 yd³ (loose)

The number of trucks needed to balance the production of the excavator, assuming 8-hr workdays, is most nearly:

A. 2
B. 3
C. 4
D. 5

617

A roller is used to compact soil to a specific density. The roller can achieve this target density for a 6-in layer of material in four passes. The following specifications are given:

Width of roller drum = 8.0 ft
Forward speed of roller = 3 mph
Overall efficiency of roller = 80%
Soil shrinkage = 15%

The material delivery rate (bank measure) to the site is 950 yd³/hr. The number of rollers required to keep up with this material delivery rate is most nearly:

A. 1
B. 2
C. 3
D. 4

618

A crane is used to lift a load of W = 40 tons as shown in the following figure. The crane cabin has a ballasted weight of 50 k. The allowable soil pressure is 3,800 lb/ft². If the crane is supported by four outriggers as shown, the maximum lateral offset X (ft) is most nearly:

A. 22.5

B. 16.3

C. 32.0

D. 27.0

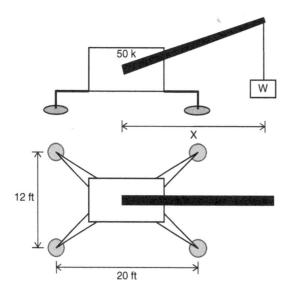

619

A boring log is shown in the following figure. Concrete piles capped by a 24-in-thick pile cap are to be driven to the rock layer. The top of the pile cap is at a depth of 3 ft. Piles are embedded 12 in into the pile cap. The required length (ft) of piles is most nearly:

A. 27

B. 28

C. 29

D. 30

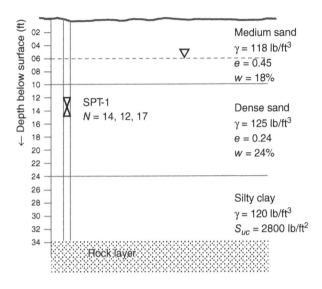

620

A certain project includes the activities summarized in the following table. The start date of the project is week 0. The project duration (weeks) is most nearly:

A. 32

B. 36

C. 41

D. 44

Activity	Predecessor Activities	Duration (Weeks)
Start	—	0
A	Start	9
B	A	8
C	B	2
D	A, C	6
E	D	7
F	C, D	9
G	E, F	7
Finish	G	0

621

As part of the PERT analysis of a project, durations are estimated for activities A through G and summarized in the following table. The critical path for the project is ACEF.

Activity	Duration (Weeks)		
	Optimistic	Most Probable	Pessimistic
A	3	4	5
B	4	5	7
C	5	7	8
D	3	4	5
E	5	6	7
F	7	9	10
G	4	5	6

The probability (%) that the project will be completed in less than 25 weeks is most nearly:

A. 12

B. 18

C. 20

D. 22

622

The activity on the node diagram for a project is shown in the following figure. All durations are in weeks. All relationships are finish to start unless noted otherwise. The free float (weeks) of activity A is:

A. 0

B. 1

C. 2

D. 3

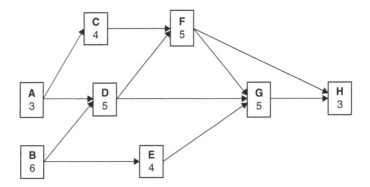

623

A construction project is broken down into seven stages (A–G). The sequencing of these stages is represented by a Gantt chart as shown in the following figure. The time axis is marked in units of (5-day) weeks. The Gantt chart shows current completion levels (shaded in gray) superimposed on the initially planned timeline. The current status of the project at 4 weeks 2 days is shown here. Which of the following statements is true?

A. All activities in stage E must be complete before the initiation of stage G.

B. The only stage which is on or ahead of schedule is G.

C. Stages A and D are behind schedule.

D. Of ongoing or completed stages, those with the lowest levels of completion are C and D.

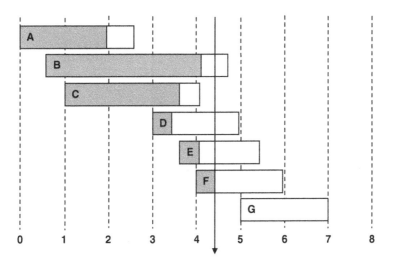

624

A new earth haul project requires moving of 323,000 ft³ (bank measure) of excavated material to a location 12 mi away. A bidding contractor had completed a similar project (size and scope) 2½ years ago. The bid price for the job on record was $525.00 per dump truck load. Annual construction inflation factor is 3.2%. The contractor uses a fleet of dump trucks with 26 yd³ capacity (heaped). The excavated material has a swell factor of 25%. The contractor's bid ($) should be most nearly:

A. 262,000
B. 302,000
C. 326,000
D. 353,000

625

A 6-in square unreinforced-concrete beam is loaded in third point loading on a simple span = 5 ft. The test is conducted 28 days after the concrete beam is cast. If the failure occurs by flexural cracking within the middle third of the beam when load P = 800 lb, the 28-day compressive strength (lb/in²) of the concrete is most nearly:

A. 6,500
B. 5,500
C. 4,500
D. 3,500

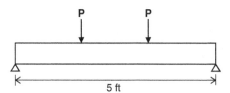

626

A 6-in-diameter × 12-in-height concrete cylinder fails at a transverse load of 55,000 lb at 28 days in a split cylinder test. The 28-day compression strength (lb/in²) of the concrete is most nearly:

A. 6,200
B. 5,200
C. 4,200
D. 3,200

627

Which of the following weld symbols is/are incorrect?

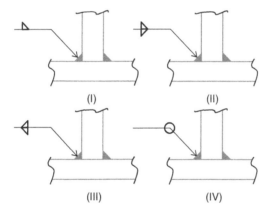

A. I and IV only
B. I and II only
C. II and III only
D. I, III, and IV only

628

A concrete mix has the following components:

Cement:	160 lb
Wet sand (moisture content = 5%):	290 lb
Wet coarse aggregate (moisture content = 3%):	420 lb
Added water:	56 lb
Air:	3% (by volume)

The following specifications are given:

Cement	specific gravity = 3.15
SSD sand	(m.c. = 0.7%) specific gravity = 2.70
SSD coarse aggregate	(m.c. = 0.5%) specific gravity = 2.60

The unit weight of the concrete (lb/ft³) is most nearly:

A. 140
B. 142
C. 144
D. 146

629

Specifications for a construction job state that wall forms can be removed once the concrete reaches a compressive strength of 3,400 psi. The maturity curve (Nurse-Saul) for the concrete is shown in the figure below. The cement hydration is assumed to cease below a temperature of 30°F. The temperature of the concrete at the time of the opening is 70°F. The number of hours that the contractor has to wait before removing forms is most nearly:

A. 24
B. 36
C. 48
D. 60

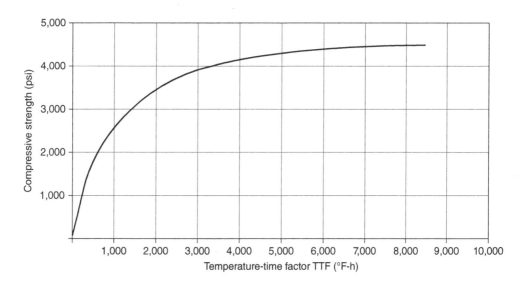

630

Concrete cylinders are tested as part of QA/QC program for a building structure. The underlying probability of individual cylinders testing at or above the target strength is 90%. If the desired confidence level is 80%, the minimum number of cylinders must test defect-free is most nearly:

A   16
B   12
C   9
D   5

631

A retaining wall (height = 24 ft) is shown in the following figure. Lateral loads are resisted by geotextile blankets spaced at 3-ft vertical spacing as shown. The angle of friction between the backfill soil and the reinforcing strips is $\delta = 15°$. The required length (ft) of the reinforcing strip at a depth D = 10 ft below the surface is most nearly:

A. 2

B. 10

C. 8

D. 16

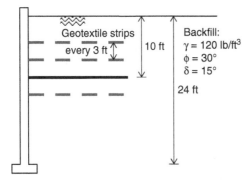

632

A 14-ft-high wall is placed at a rate of 4 ft/hr. Concrete is lightweight (135 pcf) made with Type I cement and a retarding admixture. The average concrete temperature is 60°F. The lateral design pressure exerted by the fresh concrete is most nearly:

A. 870

B. 750

C. 900

D. 725

633

According to OSHA, lifelines shall be secured above the point of operation to an anchorage or structural member capable of supporting a dead weight of at least:

A. 5,000 lb

B. 4,500 lb

C. 4,000 lb

D. 3,200 lb

634

A multistory concrete building is being constructed using one level of shoring and two levels of reshores. The dead load of a floor slab is designated D. Construction live load (due to presence of workers and equipment) is 0.3D, weight of forms and shores is 0.1D, and weight of reshores is 0.05D.

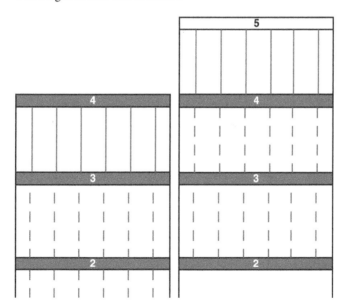

(*Continued on next page*)

At a certain stage of the project, where the last poured slab (no. 4) has hardened and is self-supporting. During the next stage of construction, the reshores below level 2 are removed and flown up to level 4 and the slab for floor 5 is poured. At the end of this stage, the load in the reshores below level 4 is:

A. 1.34D

B. 1.12D

C. 1.05D

D. 0.98D

635

Whenever a masonry wall is constructed, a limited access zone is to be established. Which of the following is true?

A. The limited access zone shall run to the same length on both sides of the wall.

B. The width of the limited access zone shall be wall height plus 2 ft.

C. The width of the limited access zone shall be wall height plus 4 ft.

D. If provided along the entire length of the wall, the width of the limited access zone is based on the judgment of the contractor.

636

During construction of an ordinary moment frame steel building, the total incident loads on a beam are as follows:

Wheeled carts carrying materials – Loaded weight of a single cart = 600 lb with a maximum of three vehicles at a time
Two workers
Total dead load = 15 k

The nominal horizontal load (lb) that should be used in the design of the beam and its connections is most nearly:

A. 100

B. 120

C. 180

D. 300

637

A retaining wall provides lateral support to a granular backfill ($\phi = 34°$) to a height of 16 ft as shown in the following figure. The backfill surface is horizontal and is used to stockpile material excavated from the site.

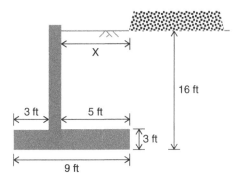

The minimum setback distance X (ft) is most nearly:

A. 8.5
B. 11.9
C. 13.5
D. 15.0

638

During an 8-hr workday, a construction worker is subjected to the following noise levels:

< 90 dB	3 hr
90 dB	3 hr
95 dB	1 hr
100 dB	1 hr

According to OSHA, the noise exposure factor for this worker is most nearly:

A. 0.89
B. 0.94
C. 1.07
D. 1.12

639

The safety log for a steel fabrication yard is as follows:

Total number of full-time employees	400
Average number of hours worked during the year	1,970
Total number of deaths	0
Total number of cases with days away from work	13 cases
Total number of days away from work	120 days
Total other recordable cases	18 cases

What is most nearly the annual total recordable case rate, as defined by OSHA?

A. 5

B. 8

C. 18

D. 31

640

A temporary road closure (not exceeding 20 min during the daytime) is established around a 1,250 ft long work zone (including a buffer space) on a freeway. The distance (ft) from the centerline of the work zone to the location of the ROAD WORK AHEAD sign is most nearly:

A. 4,800

B. 5,800

C. 6,800

D. 7,800

END OF CONSTRUCTION DEPTH EXAM

USE THE ANSWER SHEET ON THE NEXT PAGE

Construction Depth Exam: Answer Sheet

601	Ⓐ	Ⓑ	Ⓒ	Ⓓ
602	Ⓐ	Ⓑ	Ⓒ	Ⓓ
603	Ⓐ	Ⓑ	Ⓒ	Ⓓ
604	Ⓐ	Ⓑ	Ⓒ	Ⓓ
605	Ⓐ	Ⓑ	Ⓒ	Ⓓ
606	Ⓐ	Ⓑ	Ⓒ	Ⓓ
607	Ⓐ	Ⓑ	Ⓒ	Ⓓ
608	Ⓐ	Ⓑ	Ⓒ	Ⓓ
609	Ⓐ	Ⓑ	Ⓒ	Ⓓ
610	Ⓐ	Ⓑ	Ⓒ	Ⓓ
611	Ⓐ	Ⓑ	Ⓒ	Ⓓ
612	Ⓐ	Ⓑ	Ⓒ	Ⓓ
613	Ⓐ	Ⓑ	Ⓒ	Ⓓ
614	Ⓐ	Ⓑ	Ⓒ	Ⓓ
615	Ⓐ	Ⓑ	Ⓒ	Ⓓ
616	Ⓐ	Ⓑ	Ⓒ	Ⓓ
617	Ⓐ	Ⓑ	Ⓒ	Ⓓ
618	Ⓐ	Ⓑ	Ⓒ	Ⓓ
619	Ⓐ	Ⓑ	Ⓒ	Ⓓ
620	Ⓐ	Ⓑ	Ⓒ	Ⓓ

621	Ⓐ	Ⓑ	Ⓒ	Ⓓ
622	Ⓐ	Ⓑ	Ⓒ	Ⓓ
623	Ⓐ	Ⓑ	Ⓒ	Ⓓ
624	Ⓐ	Ⓑ	Ⓒ	Ⓓ
625	Ⓐ	Ⓑ	Ⓒ	Ⓓ
626	Ⓐ	Ⓑ	Ⓒ	Ⓓ
627	Ⓐ	Ⓑ	Ⓒ	Ⓓ
628	Ⓐ	Ⓑ	Ⓒ	Ⓓ
629	Ⓐ	Ⓑ	Ⓒ	Ⓓ
630	Ⓐ	Ⓑ	Ⓒ	Ⓓ
631	Ⓐ	Ⓑ	Ⓒ	Ⓓ
632	Ⓐ	Ⓑ	Ⓒ	Ⓓ
633	Ⓐ	Ⓑ	Ⓒ	Ⓓ
634	Ⓐ	Ⓑ	Ⓒ	Ⓓ
635	Ⓐ	Ⓑ	Ⓒ	Ⓓ
636	Ⓐ	Ⓑ	Ⓒ	Ⓓ
637	Ⓐ	Ⓑ	Ⓒ	Ⓓ
638	Ⓐ	Ⓑ	Ⓒ	Ⓓ
639	Ⓐ	Ⓑ	Ⓒ	Ⓓ
640	Ⓐ	Ⓑ	Ⓒ	Ⓓ

8

Breadth Exam No. 1
Solutions

These detailed solutions are for questions 1 to 40, representative of a 4-hr breadth exam according to the syllabus and guidelines for the Principles and Practice (P&P) of Civil Engineering Examination administered by the National Council of Examiners for Engineering and Surveying (NCEES), current for the October 2020 examination.

1

One set of forms will be new (area of both sides = $2 \times 12 \times 20 = 480 \text{ ft}^2$) and two sets (area = 960 ft^2) will be reused.

Volume of concrete, not including waste: $V_c = 12 \times 60 \times 1 = 720 \text{ ft}^3 = 26.67 \text{ yd}^3$

Total volume of concrete, including waste: $V_c = 1.1 \times 26.67 = 29.33 \text{ yd}^3$

Therefore, cost of erecting forms = $4.30 \times 480 + 1.30 \times 960 = \$3,312$

Cost of dismantling forms = $1.05 \times 1440 = \$1,512$

Cost of concrete (incl. waste) = $120 \times 29.33 = \$3,519.60$

Cost of reinforcement (not incl. waste) = $25 \times 26.67 = \$666.67$

Total cost = $9,011$ **(D)**

2

Crew labor rate:

$$\frac{1 \times 30 + 2 \times 18 + 1 \times 12}{4} = \$19.50/\text{L.H.}$$

Area: $A = 2 \times (35 + 25) \times 14 + 35 \times 25 - 85 = 2{,}470 \text{ ft}^2$

Labor cost:

$$2470 \text{ ft}^2 \div \frac{150 \text{ ft}^2}{LH} \times \frac{\$19.50}{LH} + 2470 \text{ ft}^2 \div \frac{50 \text{ ft}^2}{LH} \times \frac{\$19.50}{LH} = \$1{,}284.40 \qquad \text{(B)}$$

3

A has $ES_A = 0$, $EF_A = 4$

D has (single predecessor A) $ES_D = 4$, $EF_D = 11$

B has $ES_B = 0$, $EF_B = 3$

E has (single predecessor B) $ES_E = 3$, $EF_E = 9$

G has $ES_G =$ larger of $EF_D = 11$ and $EF_E = 9$

Therefore, $ES_G = 11$ weeks. $\qquad$ (C)

4

Activity on node version of the network is shown as follows. Each task is labeled with its duration in the subscript.

Solution progresses as follows (ES: early start; EF: early finish)

Activity A: $ES_A = 0$; $EF_A = 5$

Activity D: has only one predecessor (A). $ES_D = 5$; $EF_D = 9$

Activity G: FF lag = 5 with D. $EF_G = 14$

Activity J: has multiple predecessors (F, G, I), whose EF times are (not shown) 13, 14, and 13 respectively. Therefore, $ES_J = 14$; $EF_J = 16$

The critical path is ADGJ (length 16 months). Answer is $\qquad$ (B)

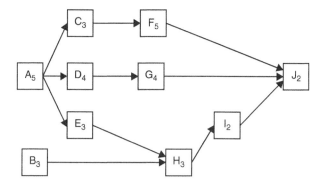

5

The cost performance index is calculated as

$$CPI = \frac{BCWP}{ACWP} = \frac{488,000}{510,000} < 1$$

Therefore, the project is currently over budget

The schedule performance index is calculated as

$$SPI = \frac{BCWP}{BCWS} = \frac{488,000}{435,000} > 1$$

Therefore, the project is currently ahead of schedule. (A)

6

Truckload = 10 tons = 20,000 lb of soil

Soil volume in each truckload = 20,000 ÷ 165 = 121.2 ft³ = 4.49 yd³

Thus, truck moves 4.49 yd³ every 30 minutes.

Thus, in a 10-hr workday, a single truck makes 600 min/30 min = 20 trips, moving 20 × 4.49 = 90 yd³.

In 8 days, a single truck moves 720 yd³

Total number of trucks needed = 4,200/720 = 5.83. Use 6 trucks. (B)

7

The solution will assume that the critical condition is overturning due to wind gusts.

Resultant wind force on the sign = $14 \times 18 \times 55 / 144 = 96.25$ lb

Height to center of sign = 41 in

Overturning moment about the tipping heel = $96.25 \times 41 = 3{,}946.3$ lb-in

Stabilizing moment = $W_{bucket} \times 7$ in = $3{,}946.3$. Therefore, $W_{bucket} = 563.8$ lb.   **(D)**

8

For horizontal backfill, and ignoring friction between the back of the wall and backfill (i.e., using Rankine's theory):

$\phi = 34° \rightarrow K_a = 0.283$

Resultant earth pressure force: $R_a = \dfrac{1}{2} \times 0.283 \times 120 \times 17^2 = 4{,}900$ lb/ft

Total gravity force (weight of concrete wall components and soil above the heel) is

$W = (11 \times 3 + 14 \times 1) \times 150 + 6 \times 14 \times 120 = 17{,}130$ lb/ft

Available friction force: $F_f = W \tan 20 = 6{,}234.8$ lb/ft

Factor of safety for sliding instability:

$FS = \dfrac{6{,}235}{4{,}900} = 1.27$ **(A)**

9

Bearing pressure applied to the soil: $q = \dfrac{P}{B^2} = \dfrac{6000}{12^2} = 41.67$ psi

Settlement: $\Delta = \dfrac{q}{k} = \dfrac{41.67}{500} = 0.083$ in   **(C)**

10

For sample A: Normal stress $\sigma_1 = 1,000$ psf

Shear stress $\tau_1 = 675$ psf

For sample B: Normal stress $\sigma_2 = 3,000$ psf

Shear stress $\tau_2 = 2,025$

Friction angle is calculated from $\tan\phi = \dfrac{\tau_2 - \tau_1}{\sigma_2 - \sigma_1} = \dfrac{2,025 - 675}{3,000 - 1,000} = 0.675 \Rightarrow \phi = 34°$

Cohesion: $c = \dfrac{\tau_1\sigma_2 - \tau_1\sigma_1}{\tau_2 - \sigma_1} = \dfrac{675 \times 3,000 - 2,025 \times 1,000}{3,000 - 1,000} = 0$ (A)

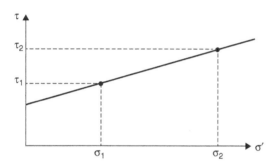

11

The following are Terzaghi's values of the bearing capacity factors or friction angle $\phi = 10°$

$N_c = 9.6; N_q = 2.7; N_\gamma = 0.56$

For a soil with small angle of internal friction, cohesion is approximately half the unconfined compression strength = 1,200 lb/ft²

For a square footing, Terzaghi's bearing capacity equation yields:

$q_{ult} = 1.3cN_c + \gamma DN_q + 0.4\gamma BN_\gamma$

(*Continued on next page*)

$$q_{ult} = 1.3 \times 1200 \times 9.6 + 122 \times 3 \times 2.7 + 0.4 \times 122 \times B \times 0.56 = 15{,}964 + 27B$$

$$q_{all} = \frac{a_{ult}}{FS} = \frac{15{,}964 + 27B}{3} = 5{,}321 + 9B$$

Allowable bearing load (lb) must be greater than the applied column load:

$$(5{,}321 + 9B)B^2 \geq 80{,}000 \Rightarrow B \geq 3.87 \text{ ft} \tag{C}$$

12

For sample 5, field N value $= 11 + 14 = 25$

Sample 5 extends from a depth of 13.5 ft to 15.0 ft. The depth corresponding to the effective center of the sample is 14.5 ft.

GWT is at a depth of 8 ft.

At 14.5 ft depth, effective vertical stress due to 6 ft of unsubmerged clay + 2 ft of unsubmerged sand + 6.5 ft of submerged sand:

$$\sigma'_v = 130 \times 6 + 120 \times 2 + (120 - 62.4) \times 6.5 = 1394.4 \, psf = 0.7 \, tsf$$

Overburden correction factor (Liao & Whitman): $C_N = \sqrt{\dfrac{1}{0.7}} = 1.2$

Corrected N value $= 1.2 \times 25 = 30$. $\hspace{3cm}$ (C)

13

Factor of safety is given by:

$$FS = \frac{c}{\gamma H \cos^2 \beta \tan \beta} + \frac{\tan \phi}{\tan \beta}$$

For the riprap layer:

$$FS = 0 + \frac{\tan 36}{\tan 30} = 1.26$$

For the soil:

$$FS = \frac{600}{125 \times 20 \times \cos^2 30 \tan 30} + \frac{\tan 28}{\tan 30} = 0.55 + 0.92 = 1.47$$

The governing FS is 1.26 $\hspace{3cm}$ (A)

14

Using lever rule, the reaction on the left support: $R_A = 42 \times \dfrac{13.5}{20} + 20 \times \dfrac{5}{20} = 33.35$

At the beginning of the distributed load ($x = 3$), the shear $= 33.35$

Since the distributed load is applied at 6 k/ft, it will take $33.35/6 = 5.56$ ft for the shear $= 0$

Measured from the left support, $X = 8.56$ ft. (B)

15

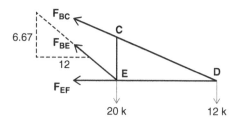

Making a section through BC, BE, and EF and then taking moments about D (members BC and EF pass through point D, and they will be absent from this equation)

$$\sum M_D = \frac{6.67}{13.73} F_{BE} \times 12 - 20 \times 12 = 0 \Rightarrow F_{BE} = +41.17 \ k$$

Answer is (D)

16

At the support, the internal forces are:

Shear force: $V = 10 \sin 40 = 6.428$ k

Axial force: $A = 10 \cos 40 = 7.660$ k

Bending moment: $M = 10 \sin(40) \times 60 + 10 \cos(40) \times 20 = 538.9$ k-in (clockwise)

Since the axial force is tensile, the maximum stress (tension) at the top edge is:

$$\sigma = \frac{P}{A} + \frac{My}{I} = \frac{10}{6 \times 8} + \frac{538.9 \times 4}{\dfrac{1}{12} \times 6 \times 8^3} = 0.21 + 8.42 = 8.63 \tag{A}$$

17

Writing the equilibrium equations at C:

$$\sum F_x = 0 \Rightarrow \frac{3}{\sqrt{10}}T_2 = \frac{1}{\sqrt{5}}T_3 \Rightarrow 0.9487T_2 = 0.4472T_3$$

$$\sum F_y = 0 \Rightarrow \frac{1}{\sqrt{10}}T_2 + \frac{2}{\sqrt{5}}T_3 = 3 \text{ tons} \Rightarrow 0.3162T_2 + 0.8944 \times 2.1214T_2$$
$$= 3 \text{ tons} \Rightarrow T_2 = 1.355 \text{ tons}$$

Writing the equilibrium equations at B:

$$\sum F_x = 0 \Rightarrow \frac{1}{\sqrt{2}}T_1 = \frac{3}{\sqrt{10}}T_2 \Rightarrow 0.7071T_1 = 0.9487T_2 \Rightarrow T_1 = 1.818 \text{ tons}$$

$$\sum F_y = 0 \Rightarrow \frac{1}{\sqrt{2}}T_1 - \frac{1}{\sqrt{10}}T_2 - F = 0 \Rightarrow F = 0.7071T_1 - 0.3162T_2$$
$$= 0.86 \text{ tons} = 1,714 \text{ lb}$$ (B)

18

The temperature range causing expansion of the girders is from 65°F to 95°F. The expansion of the girder due to this $\Delta T = 30°F$, is:

$$\Delta L_T = \alpha L \Delta T = 7.3 \times 10^{-6} \times 90 \times 30 = 0.0197 \text{ ft} = 0.24 \text{ in.}$$ (C)

19

Maximum moment: $M = \dfrac{wL^2}{8} = \dfrac{2 \cdot 75 \times 25^2}{8} = 214.8 \text{ k·ft} = 2578 \text{ k-in}$

Required section modulus: $S_{reqd} = \dfrac{M}{\sigma_{all}} = \dfrac{2578}{32} = 80.6 \text{ in}^3 \rightarrow W12 \times 65$ (D)

20

Wetted perimeter: $P = 12 + 2 \times 4.6 = 21.2 \text{ ft}$

Flow area: $A = 12 \times 4.6 = 55.2 \text{ ft}^2$

Hydraulic radius: $R = \dfrac{A}{P} = \dfrac{55.2}{21.2} = 2.60$ ft

According to Chezy-Manning, average velocity: $V = \dfrac{1.486}{0.014} \times 2.6^{2/3} \times \sqrt{0.004} = 12.69$ fps

Flow rate: $Q = VA = 12.69 \times 55.2 = 700$ cfs

Alternate solution (using tables for flow parameter K – *All In One* Table 303.3)

For depth ratio $\dfrac{d}{b} = \dfrac{4 \cdot 6}{12} = 0.383 \rightarrow K = 0.1385$ (Table 303.3)

Flow rate: $Q = \dfrac{1.486}{0.014} \times 0.1385 \times 12^{8/3} \times \sqrt{0.004} = 701.8$ cfs (Equation 303.36)   **(B)**

21

The head behind the weir = $127.50 - 124.72 = 2.78$ ft

The discharge through the weir (assuming weir coefficient = 3.33) is

$Q = CbH^{3/2} = 3.33 \times 5 \times 2.78^{3/2} = 77.2$ cfs $= 49.9$ MGD   **(A)**

22

The question directs us to use the Rational Method, even though it is usually not used for areas larger than about 100 acres.

Time of concentration = longest of all overland flow times = 45 min

From the Intensity-Duration-Frequency curves, for a 20-year storm, with duration = 45 min, we get intensity $I = 3.7$ in/hr.

The composite Rational C coefficient is given by:

$$\bar{C} = \frac{\sum C_i A_i}{\sum A_i} = \frac{0.4 \times 80 + 0.2 \times 80 + 0.9 \times 50 + 0.6 \times 90 + 0.2 \times 70}{370} = 0.44$$

Rational method runoff discharge:

$$Q = CiA = 0.44 \times 3.7 \times 370 = 603 \text{ ac} \cdot \frac{\text{in}}{\text{hr}} = 608 \text{ ft}^3/\text{sec} \qquad \textbf{(B)}$$

23

The flow rate at the end of the second hour is calculated as the superposition of the effect of the first hour (with a 2-hr lag, since the first hour of excess precipitation STARTS at $t = 0$) and the effect of the second hour (with a 1-hr lag). Each of these ordinates is scaled by the appropriate depth of excess precipitation (1.7 in and 0.8 in, respectively)

$$Q_3 = 95 \times 1.7 + 30 \times 0.8 = 185.5 \text{ cfs} \tag{D}$$

24

Rain Gage Station ID	Area (A)	Depth D (in)	A×D
A-1	2.1	1.7	3.57
A-2	3.2	2.1	6.72
A-3	5.2	1.9	9.88
B-1	2.9	1.2	3.48
B-2	4.9	1.1	5.39
B-3	6.1	0.9	5.49
B-4	3.1	1.5	4.65
B-5	2.2	2.0	4.40
	29.7		43.58

The average precipitation depth (in) is given by the weighted average:

$$\bar{P} = \frac{\sum D_i A_i}{\sum A_i} = \frac{43.58}{29.7} = 1.47 \tag{C}$$

25

Flow rate $Q = 950 \text{ gpm} = 2.12 \text{ cfs}$

Ignoring energy loss in the reducer,

$$z_1 + \frac{V_1^2}{2g} + \frac{p_1}{\gamma} = z_2 + \frac{V_2^2}{2g} + \frac{p_2}{\gamma} \Rightarrow \frac{p_1 - p_2}{\gamma} = \frac{V_2^2 - V_1^2}{2g} + z_2 - z_1$$

Since the pipe is horizontal, $z_1 = z_2$

Velocity in the upstream section: $V_1 = \dfrac{Q}{A} = \dfrac{2.12}{\pi\left(\dfrac{6}{12}\right)^2} = \dfrac{2.12}{0.196} = 10.8$ fps

By continuity, since the area ratio is 4:1, the velocity ratio must be 1:4. So, $V_2 = 43.1$ fps

$\dfrac{p_1 - p_2}{\gamma} = \dfrac{43.1^2 - 10.8^2}{2 \times 32.2} + 0 = 27.1$ ft

Since 1 atm = 33.9 ft H_2O = 14.7 psi, a head loss of 27.1 ft is equivalent to a pressure loss of 11.75 psi. **(D)**

26

Since flow occurs by gravity alone, application of Bernoulli's equation between the two reservoir surfaces will yield:

$$z_1 + \frac{V_1^2}{2g} + \frac{p_1}{\gamma} - h_f = z_2 + \frac{V_2^2}{2g} + \frac{p_2}{\gamma}$$

Since the two reservoir surfaces have negligible velocity and both are at atmospheric pressure,

$V_1 = V_2 \approx 0 \;\&\; p_1 = p_2 = p_{atm}$

$h_f = z_2 - z_1 = 70$ ft

Equivalent length of pipe: $L_{eq} = 2{,}500 + 55 = 2{,}555$ ft

$h_f = 70 = f\dfrac{L}{D}\dfrac{V^2}{2g} = 0.02 \times \dfrac{2{,}555}{2} \times \dfrac{V^2}{64.4} \Rightarrow V = 13.3$ fps

Flow rate: $Q = V \times A = 13.3 \times \dfrac{\pi}{4} \times 2^2 = 41.73$ cfs **(A)**

27

The radius of the curve is: $R = \dfrac{5{,}729.578}{D} = \dfrac{5{,}729.578}{4} = 1{,}432.4$ ft

The tangent length is the distance from PC to PI:

$T = \sqrt{(1{,}232.56 - 509.72)^2 + (123.32 + 172.11)^2} = 780.88$ ft

(*Continued on next page*)

The azimuth of the back tangent is calculated from: $\tan Az = \dfrac{\Delta E}{\Delta N} = \dfrac{-172.11 - 123.32}{509.72 - 1{,}232.56} \Rightarrow$

$Az = 202.23°$ (third quadrant angle since both ΔE and ΔN are negative)

Deflection angle: $I = 2 \tan^{-1}\left(\dfrac{T}{R}\right) = 57.2°$ to the left (counterclockwise)

Therefore, azimuth of forward tangent $= 202.23 - 57.2 = 145.03$

Change in coordinates from PI to PT (using tangent length $T = 780.88$ ft)

$\Delta N = T \cos Az = 780.88 \times \cos 145.03 = -639.90$ ft

$\Delta E = T \sin Az = 780.88 \times \sin 145.03 = +447.56$ ft

Coordinates of the PT: $(509.72 - 639.90, -172.11 + 447.56)$ or $(130.18\ \text{S},\ 275.45\ \text{E})$ **(D)**

28

The tangent offset at any location on a vertical curve $= \frac{1}{2} Rx^2$

At the end of the curve (i.e., at the PVT) $x = L$, therefore tangent offset $= \dfrac{1}{2}\dfrac{G_2 - G_1}{L}L^2 = \dfrac{(G_2 - G_1)L}{2}$

For a crest curve, the vertical offset is negative, therefore: $\dfrac{(-4 - 5)L}{2} = -17.65 \Rightarrow L = 3.9222$ sta

Therefore, since the PVC is half the curve length upstream of the PVI:

Sta. PVC = sta. PVI $- 1.9611 = 123.325 - 1.9611 = 121.3639\ (121 + 36.39)$ **(D)**

29

Pedestrians using the walkway in the first hour $= 0.9 \times 0.3 \times 40{,}000 = 10{,}800$ ped/hr (hourly average)

Peak flow during the first hour $= 10{,}800 \div 0.88 = 12{,}273$ ped/hr $= 204.5$ ped/min

Peak flow rate $= 204.5 \div 32 = 6.4$ ped/min/ft **(C)**

30

Opening size for no. 200 sieve = 0.075 mm. From the particle size distribution curve, the % passing corresponding to this size is $F_{200} = 12\%$. Since this is less than 50%, soil is coarse grained. First letter is S or G. This eliminates choices C and D. The other two choices are both "sand" (first letter S), therefore, *given these answer choices*, we do not need to distinguish between G or S.

Percentage passing no. 4 sieve (sieve size 4.75 mm), $F_4 = 98\%$. % retained $R_4 = 2\%$. This is less than half of the coarse fraction = $100 - 12 = 88\%$. Soil is predominantly sand (not gravel). First letter is S.

The three points marked on the gradation curve correspond to 10%, 30%, and 60% passing.

$D_{10} = 0.065$ mm, $D_{30} = 0.18$ mm, $D_{60} = 0.5$ mm

$$C_u = \frac{D_{60}}{D_{10}} = \frac{0.5}{0.065} = 7.7 \quad \text{and} \quad C_z = \frac{D_{30}^2}{D_{10}D_{60}} = \frac{0.18^2}{0.065 \times 0.5} = 1.0$$

Criteria for well graded are approximately met. The only available choice that fits is SW.   (A)

31

Soil sample unit weight: $\gamma = \dfrac{3.64}{0.031} = 117.4 \, \text{lb/ft}^3$

The volume of water added until all voids are filled = 5.6 oz. This must be the volume of air in the original sample.

$V_{air} = 5.6$ oz $= 0.00585 \, \text{ft}^3$

Therefore, the remaining volume is occupied by soil solids and water

$V_s + V_w = 0.031 - 0.00585 = 0.02515 \, \text{ft}^3$ (a)

Since these two constituents contribute to the soil weight, we write the expression for sample weight:

$2.65 \times 62.4 \times V_s + 62.4 \times V_w = 3.64$ (b)

If we solve equations (a) and (b), we get: $V_s = 0.020 \, \text{ft}^3$

Weight of solids: $W_s = 3.326 \, \text{lb}$

Dry unit weight of the soil: $\gamma_d = 3.326 \div 0.031 = 107.3 \, \text{lb/ft}^3$ (C)

32

For each sack of cement (weight = 94 lb)

Sand (SSD): Weight = $1.8 \times 94 = 169.2$ lb (this represents sand with 0.5% moisture)

When wet sand (m.c. = 6%) is used, it contains free water = $5.5/100.5 \times 169.2 = 9.26$ lb

Coarse aggregate (SSD): W = $2.6 \times 94 = 244.4$ lb (this represents aggregate with 0.7% moisture)

When wet aggregate (m.c. = 4%) is used, it contains free water = $3.3/100.7 \times 244.4 = 8.01$ lb

Total extra water = $9.26 + 8.01 = 17.27$ lb = 2.07 gal (8.3454 lb/gal H_2O)

Total water = $5.8 + 2.07 = 7.87$ gal/sack **(C)**

33

Yield stress = 36 ksi

Modulus of elasticity of steel = 29,000 ksi

Yield strain: $\varepsilon_y = \dfrac{\sigma_y}{E} = \dfrac{36}{29{,}000} = 0.00124 = 0.124\%$ **(B)**

34

By using the 0.2% offset method (drawing a line parallel to the initial tangent through the strain offset = 0.2% = 2000 $\mu\varepsilon$), the yield stress is 38 ksi.

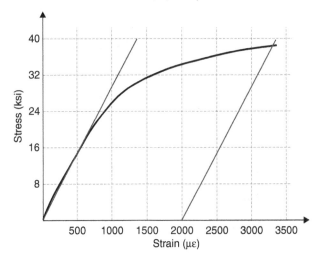

The cross section of the coupon (in the test section) is $A = 0.5 \times 0.25 = 0.125$ in^2

Axial force causing yield (at 38 ksi) $= 38{,}000 \times 0.125 = 4{,}750$ lb (B)

35

The volume of the standard Proctor mold is $1/30$ ft^3. This volume is used to reduce the net soil mass to wet unit weight. For example, for sample 1: 3.2 lb $\div$ $(1/30) = 96$ lb/ft^3

The total unit weight is then converted to a dry unit weight using $\gamma_d = \gamma/(1+w) =$ $96/1.128 = 85.1$

Weight of Soil (lb)	Water Content (%)	Unit Weight (pcf)	Dry Unit Weight (pcf)
3.20	12.8	96.0	85.1
3.78	13.9	113.4	99.6
4.40	15.0	132.0	114.8
4.10	15.7	123.0	106.3
3.70	16.6	111.0	95.2
3.30	18.1	99.0	83.9

The maximum dry unit weight $= 114.8$ lb/ft^3 (D)

36

Using the average end-area method, we calculate the total volume of cut between stations $0 + 0.00$ and $4 + 0.00$ as:

$$V_C = \frac{100}{2}[245.0 + 546.2 + 2 \times (312.5 + 111.5 + 234.5)] = 105{,}410 \text{ ft}^3$$

Similarly, the unadjusted volume of fill between stations $0 + 0$ and $4 + 0$ is:

$$V_F = \frac{100}{2}[423.5 + 514.5 + 2 \times (176.3 + 303.0 + 188.4)] = -113{,}670 \text{ ft}^3$$

Adjusting the fill volume for shrinkage (shrinkage factor $= 0.85$), we have

$V_F = -113{,}670 \div 0.85 = -133{,}729.4$ ft^3

Net earthwork between $0 + 0.00$ and $4 + 0.00 = 105{,}410 - 133{,}729.4 = -28{,}319.4$ ft$^3 =$ -1048.87 yd^3

Mass diagram ordinate at station $4 + 0.00$ is $+400 - 1048.87 = -648.87$ (A)

37

The information on the stake indicates that this is a drainage stake that identifies plan feature 13-A (option B), which is a grate whose centerline is offset 15.35 ft from the stake (option C). The elevation 823.23 refers to the grade elevation at the stake location (and not at the flow line invert, as option A indicates). The only choice that is incorrect is   **(A)**

38

Minimum length 50 ft. Minimum width 10 ft. 2–3 in stone 6 in deep over geotextile fabric. Answer is   **(D)**

39

Underpinning (I) is commonly used to support existing structures to counter the possible loss of bearing support from adjacent excavation. Slurry walls (III) can be used to "isolate" sensitive structures from construction activities.   **(B)**

40

At the rate of return (i), the present worth should be zero.

The $350k capital expenditure is a present value (P)   NEGATIVE

The $25k reduction in annual costs is an annuity (A)   POSITIVE

The $200k increase in salvage value is a future sum (F)   POSITIVE

Converting all of these to present worth, the net present worth can be written:

$$PW = -350 + 25\left(\frac{P}{A}, i, 20 \text{ yrs}\right) + 200\left(\frac{P}{F}, i, 20 \text{ yrs}\right) = 0$$

For $i = 5\%$, PW = 37k

For $i = 6\%$, PW = -0.9k   Actual answer 5.97%

Answer is   **(C)**

Answer Key for Breadth Exam No. 1

1	D
2	B
3	C
4	B
5	A
6	B
7	D
8	A
9	C
10	A

11	C
12	C
13	A
14	B
15	D
16	A
17	B
18	C
19	D
20	B

21	A
22	B
23	D
24	C
25	D
26	A
27	D
28	D
29	C
30	A

31	C
32	C
33	B
34	B
35	D
36	A
37	A
38	D
39	B
40	C

9

Breadth Exam No. 2
Solutions

These detailed solutions are for questions 101 to 140, representative of a 4-hr breadth exam according to the syllabus and guidelines for the Principles and Practice (P&P) of Civil Engineering Examination administered by the National Council of Examiners for Engineering and Surveying (NCEES), current for the October 2020 examination.

101

Excavation for grade beam has a bottom width = 3 ft, 1:1 side slopes and a depth of 2 ft, which creates a top width: $T = 3 + 2 \times 2 = 7$ ft

Cross section of trench: $A = \dfrac{1}{2} \times (3 + 7) \times 2 = 10 \text{ ft}^2$

Perimeter of building: $P = 200 \times 2 + 400 \times 2 + 50 \times 2 = 1{,}300$ ft

Volume of excavation: $V = 10 \times 1{,}300 = 13{,}000 \text{ ft}^3 = 481.5 \text{ yd}^3$

Daily productivity $= 9 \dfrac{\text{yd}^3}{\text{hr}} \times 8 \dfrac{\text{hr}}{\text{day}} = 72 \dfrac{\text{yd}^3}{\text{day}}$

Number of days = 481.5 ÷ 72 = 6.7 days. Choose 7 days as minimum number of days to complete job. **(D)**

102

AD	11 weeks	8,000
AE	6 weeks	6,000
BE	7 weeks	7,000
CF	10 weeks	9,000

To complete the project in 9 weeks, AD must be shortened by 2 weeks (shorten D by 2 weeks for additional $1,000) AND CF must be shortened by 1 week (shorten C by 2 weeks for additional $3,000). Bonus due to 2 weeks early completion is $2,000.

Net revised cost = original cost + extra cost − bonus = 24k + 1k + 3k − 2k = $26,000   (D)

103

The following diagram shows forward pass through the network. Time to complete = 20   (B)

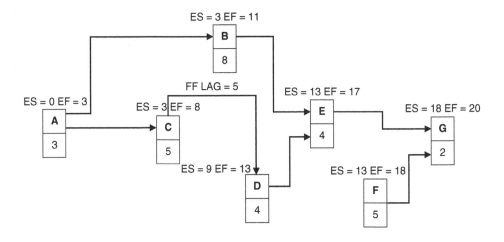

104

Calcium chloride, being hygroscopic as well as resistant to evaporation, is a common choice as a dust control measure on construction sites.   (A)

105

Since the outriggers are simply resting on the soil (not anchored), the limiting condition is when the far side (left) outrigger legs have zero reaction. For this condition, the inside legs carry a total load of 40 tons. (Incidentally, the offset X for which this occurs is 15 ft from moment equilibrium.)

Therefore, maximum load per leg = 20 tons = 40,000 lb. Based on the allowable soil pressure of 2,800 psf, the outrigger pads must have a minimum area = 40/2.8 = 14.3 ft² (each). In this case, choose a bigger pad. (C)

106

Waste produced by the original population of town A = 20,000 × 5 = 100,000 lb/day

At a compacted density of 40 lb/ft³, the volume taken up in the landfill = 100,000 ÷ 40 = 2,500 ft³/day = 912,500 ft³/yr = 33,796 yd³/yr

Remaining capacity = 1 × 10⁶/33,796 = 29.6 years

Waste produced by expanded population of town A = 25,000 × 5 = 125,000 lb/day

At a compacted density of 40 lb/ft³, the volume taken up in the landfill = 125,000 ÷ 40 = 3,125 ft³/day = 1,140,625 ft³/yr = 42,245 yd³/yr

Remaining capacity = 1 × 10⁶/42,245 = 23.7 years

Reduction in service life = 29.6 − 23.7 = 5.9 years (D)

107

The weight of the wall panel (14′H × 20′L × 4″ T) = 80 × 14 × 20 × 4/12 = 7,467 lb

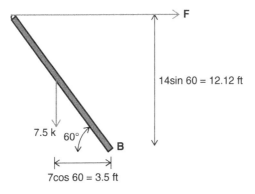

F

14sin 60 = 12.12 ft

7.5 k 60°

B

7cos 60 = 3.5 ft

(Continued on next page)

Using moment equilibrium about pivot B:

$$M_B = 0 \Rightarrow 7.5 \times 3.5 - 12.12F = 0 \Rightarrow F = 2.16\,k \qquad \text{(A)}$$

108

Ignoring friction between backfill and the wall (Rankine theory), the active earth pressure coefficient is given by (simplified equation applies for $\theta = 90$, $\delta = 0$, $\beta = 0$):

$$K_a = \frac{1 - \sin 31}{1 + \sin 31} = 0.32$$

Active earth pressure resultant is given by:

$$R_a = \frac{1}{2}K_a\gamma H^2 = 0.5 \times 0.32 \times 118.4 \times 15^2 = 4{,}262.4\,\frac{\text{lb}}{\text{ft}} \qquad \text{(C)}$$

109

$$s = \frac{HC_c}{1 + e_o}\log_{10}\frac{p_2'}{p_1'} = \frac{32 \times 12 \times 0.20}{1 + 0.6}\log_{10} 2 = 14.5 \text{ in.} \qquad \text{(A)}$$

110

Tributary width S incident to each strut creates a distributed load 28S on the wall, as shown here. The support condition at the bottom of the wall can be assumed to be like a hinge (laterally restrained, rotationally unrestrained). Each strut force (F_s) is limited to 2,350 lb. The horizontal component (F_H) is therefore $0.6 \times 2{,}350 = 1{,}410$ lb. Taking moments about the bottom of the wall segment,

$$M_o = (28S)10 \times 5 - 1410 \times 8 = 0 \Rightarrow S \le 8.06 \text{ ft} \qquad \text{(C)}$$

111

Ultimate bearing capacity of a square footing, according to Terzaghi's theory, is given by

$$q_{ult} = 1.3cN_c + \gamma DN_q + 0.4\gamma BN_\gamma$$

From the supplied figure, for $\phi = 30°$, $N_c = 30$, $N_q = 18.5$, $N_\gamma = 22.5$

$$q_{ult} = 1.3 \times 200 \times 30 + 120 \times 3 \times 18.5 + 0.4 \times 120 \times 5 \times 22.5 = 19,860 \text{ psf}$$

Soil pressure at base of footing = column load + soil overburden = $140,000/25 + 120 \times 3 = 5,960$ psf

$FS = 19,860/5,960 = 3.33$ (D)

112

Deviatoric stress: $\Delta\sigma = \dfrac{95}{\dfrac{\pi}{4}(2)^2} = 30.24$ psi

Total axial stress = $30.24 + 15 = 45.24$ psi

Effective axial stress = $45.24 - 6.5 = 38.74$ psi

Total lateral stress = 15 psi

Effective lateral stress = $15 - 6.5 = 8.5$ psi

Diameter of Mohr's circle = $38.74 - 8.5 = 30.24$ psi

The cohesion is given by the radius of the circle = 15.12 psi = 2.177 psf (B)

113

The correct pattern is the one which has steel in the areas developing tensile stress due to the wall being loaded laterally (to the left). Answer is   **(B)**

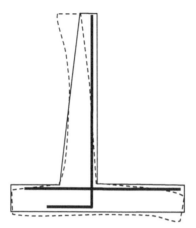

114

A section through members CD, DI, and HI is shown in the following figure. To solve for F_{DI}, the most efficient method may be to take moments about F (where the other 2 forces intersect). Note that this means that calculating the support reaction at F is also unnecessary. Height DH = 2/3 × 6 = 4 ft.

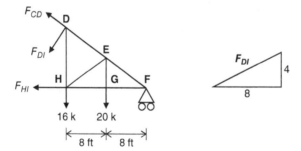

Taking moments about F for the section shown:

$$F_{DI} \cdot \frac{1}{\sqrt{5}} \times 16 + F_{DI} \cdot \frac{2}{\sqrt{5}} \times 4 + 16 \times 16 + 20 \times 8 = 0$$

This leads to:

$$F_{DI} = -38.76 \text{ k (COMPRESSION)} \tag{C}$$

115

Elastic neutral axis (centroid) location (using bottom edge of section as datum)

$$\bar{y} = \frac{\sum y_i A_i}{\sum A_i} = \frac{0.5 \times 8 + 6 \times 10 + 11.25 \times 3}{8 + 10 + 3} = 4.655$$

Using the Parallel Axis Theorem, the moment of inertia I_{xc} is given by

$$I_{xc} = \frac{1}{12} \times 6 \times 0.5^3 + 3 \times (11.25 - 4.655)^2 + \frac{1}{12} \times 1 \times 10^3 + 10 \times (6 - 4.655)^2$$
$$+ \frac{1}{12} \times 8 \times 1^3 + 8 \times (0.5 - 4.655)^2 = 370.74 \text{ in}^4$$

The farthest fiber from the neutral axis is at a distance $c = 11.5 - 4.655 = 6.845$ in

Elastic section modulus: $S_x = \dfrac{I_{xc}}{c} = \dfrac{370.74}{6.845} = 54.2 \text{ in}^3$ \hspace{1em} (A)

116

The tributary width for each joist is the center-to-center distance $= 3$ ft

Floor load incident to each joist $= 90 \text{ lb/ft}^2 \times 3 \text{ ft} = 270 \text{ lb/ft}$

Maximum bending moment in simply supported joists: $M_{max} = \dfrac{wL^2}{8} = \dfrac{270 \times 18^2}{8} =$ 10,935 lb · ft $= 131,220$ lb-in

The allowable bending stress $= 1,700$ psi

Required (minimum) section modulus: $S > \dfrac{M}{\sigma_{all}} = \dfrac{131,220}{1,700} = 77.2 \text{ in}^3$ (choose 80 in^3) \hspace{0.5em} (C)

117

For a simple span carrying a UDL, the maximum elastic deflection is given by:

$$\Delta_{max} = \frac{5wL^4}{384EI} \le \frac{L}{360} \Rightarrow I \ge \frac{1800wL^3}{384E}$$

Using compatible units—$w = 4.75$ k/ft $= 0.396$ k/in; $L = 65$ ft $= 780$ in; $E = 29,000$ k/in²

$I > 30,362$ in⁴ (B)

118

The free body diagram of joint B is shown below. Since this FBD has two unknowns, the two equations of equilibrium may be used to solve for them.

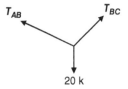

$$\sum F_x = -T_{AB}\frac{3}{\sqrt{10}} + T_{BC}\frac{1}{\sqrt{2}} = 0 \Rightarrow T_{AB} = 0.745T_{BC}$$

$$\sum F_y = T_{AB}\frac{1}{\sqrt{10}} + T_{BC}\frac{1}{\sqrt{2}} - 20 = 0$$

$$\Rightarrow \left(0.745\frac{1}{\sqrt{10}} + \frac{1}{\sqrt{2}}\right)T_{BC} = 20 \Rightarrow T_{BC} = 21.21\ k, T_{AB} = 15.81\ k \quad \text{(A)}$$

Alternate solution (since it is 3 forces in equilibrium, the force polygon is a triangle. Use the law of sines)

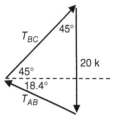

$$\frac{T_{AB}}{\sin 45} = \frac{20}{\sin 63.4} \Rightarrow T_{AB} = 15.82\ k$$

119

Modulus of elasticity of Douglas Fir-Larch = 1,500 ksi

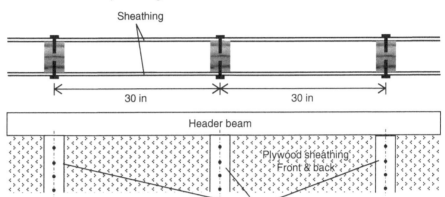

Buckling about the weak axis is prevented by the nails.

Radius of gyration about the strong axis = $0.29 \times 3.25 = 0.94$ in.

Assuming both ends to behave like "pinned" connections, $KL/r = 1.0 \times 126/0.94 = 133.7$

Euler buckling stress = $\sigma_E = \dfrac{\pi^2 E}{\left(\dfrac{KL}{r}\right)^2} = \dfrac{\pi^2 \times 1.5 \times 10^6}{133.7^2} = 828.3$ psi

Euler buckling load = $P_E = 828.3 \times 1.75 \times 3.25 = 4{,}711$ lb   **(D)**

120

The longitudinal slope is calculated as the invert elevation difference divided by the length of the pipe:

$$S = \frac{\Delta z}{L} = \frac{275.64 - 270.96}{800} = 0.00585$$

Flow rate in circular pipe flowing full (***All In One*** Equation 303.34)

$$Q_f = \frac{0.312 \times 1.486}{n} D^{8/3} S^{1/2} = \frac{0.464}{0.013} \times 2.5^{8/3} \times \sqrt{0.00585} = 31.43 \text{ cfs}$$

Using Table 303.2 from the *All In One*

Flow ratio: $\dfrac{Q}{Q_f} = \dfrac{20}{31 \cdot 43} = 0.64 \rightarrow \dfrac{d}{D} = 0.58 \rightarrow d = 17.4''$ **(C)**

121

Longitudinal slope $S = 0.005$

Side slope parameter, $m = 3$

Manning's roughness coefficient, $n = 0.020$

Bottom width, $b = 2$ ft

Depth of flow, $d = 2$ ft. Therefore, horizontal flare (each side) $= 3 \times 2 = 6$ ft

Inclined flare (each side) $= \sqrt{(2^2 + 6^2)} = \sqrt{40} = 6.325$ ft

Wetted perimeter, $P = 2 + 2\sqrt{40} = 14.65$ ft

Top width $= 14$ ft

Flow area, $A = \dfrac{2 + 14}{2} \times 2 = 16 \text{ ft}^2$

Hydraulic radius $= 16 \div 14.65 = 1.09$ ft

$$V = \frac{1.486}{0.02} \times 1.09^{2/3} \times 0.005^{1/2} = 5.56 \text{ ft/sec} \tag{A}$$

Alternatively, using Table 303.3 in *All In One PE Exam Guide* (Goswami), for $d/b = 2/2 = 1.0$ and $m = 3$, the parameter $K = 2.6725$. Using this parameter value in equation 303.36

$$\text{Flow rate, } Q = \frac{2.6725 \times 1.486 \times 2^{8/3} \times 0.005^{1/2}}{0.020} = 89.15 \text{ cfs}$$

$$\text{Flow velocity, } V = \frac{Q}{A} = \frac{89.15}{16} = 5.57 \text{ ft/sec}$$

This would require a calculation of the flow area only.

122

$$\text{Flow rate parameter: } K = \frac{Qn}{1.486 b^{8/3} S^{1/2}} = \frac{1200 \times 0.015}{1.486 \times 10^{8/3} 0.01^{1/2}} = 0.2610$$

For this K and for $m = 0$ (rectangular channel), $d/b = 0.616$

Depth of flow = $0.616 \times 10 = 6.16$ ft (for a rectangular channel, this is also hydraulic depth)

Velocity: $V = \dfrac{Q}{A} = \dfrac{1200}{10 \times 6.16} = 19.48$ fps

Froude number: $Fr = \dfrac{V}{\sqrt{gd_h}} = \dfrac{19.48}{\sqrt{32.2 \times 6.16}} = 1.38$

Answer is   (B)

123

The 1-hr unit hydrograph is used to construct the runoff contribution of the first hour and the second hour (staggered).

Time (hr)	0	1	2	3	4	5
Discharge Q (cfs/in)	0	35	75	105	40	0

Time (hr)	0	1	2	3	4	5	6
Hour 1 (scaled × 1.7)	0	59.5	127.5	178.5	68	0	
Hour 2 (scaled × 0.8)		0	28	60	84	32	0
Total	0	59.5	155.5	238.5	152	32	0

Peak discharge = 238.5 cfs   (B)

124

As the return period increases, the annual probability of occurrence decreases. This corresponds to a stronger event (higher flood elevation). I is correct.

Annual probability of 20-year storm = $1/20 = 0.05$, while annual probability of 10-year storm = $1/10 = 0.10$. So, II is correct.

III is incorrect because even though it is small, there still exists a non-zero probability that a 50-year flood will NOT occur in the next 100 years.

IV is incorrect because the magnitude of the design event is directly influenced by the return period.

Answer is   (D)

125

The composite Rational coefficient is:

$$\bar{C} = \frac{\sum C_i A_i}{\sum A_i} = \frac{0.2 \times 50 + 0.55 \times 65 + 0.85 \times 20 + 0.3 \times 240}{375} = 0.36$$

Governing time of concentration = 40 minutes. From the I-D-F curves, for a return period = 50 years, the design intensity = 1.4 in/hr

Discharge; $Q = CiA = 0.36 \times 1.4 \times 375 = 189 \, ac - \dfrac{in}{hr} = 190.5 \, ft^3/sec$ (C)

126

The slope of the pipe is 0.01. Elevation difference between the ends of the pipe = 125 – 95 = 30 ft. Therefore, the length of the pipe = 30 ÷ 0.01 = 3000 ft.

Hazen Williams, C = 120

If flow velocity is V (ft/sec), head loss due to friction is given by:

$$h_f = \frac{3.022 \times V^{1.85} \times 3,000}{120^{1.85} \times 1.0^{1.165}} = 1.291V^{1.85}$$

Bernoulli's equation applied between free surface at reservoir and free flow at the outfall:

$$145 + \frac{p_{atm}}{\gamma} + 0 - 1.29 \, V^{1.85} = 95 + \frac{p_{atm}}{\gamma} + \frac{V^2}{2g} \Rightarrow 0.0155 \, V^2 + 1.29 \, V^{1.85} = 50$$

Solving approximately (trial and error): $V = 7.15$ ft/sec

Alternatively, if the V^2 term is approximated by $V^{1.85}$, $V = 7.17$ ft/sec (without trial and error)

Flow rate, $Q = 7.15 \times \dfrac{\pi}{4} \times 1^2 = 5.62$ cfs = 2,516 gal/min   (D)

127

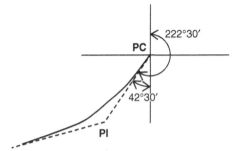

Back tangent azimuth $42.5 + 180 = 222.5$ (measured clockwise from North meridian)

$$R = 1030 \text{ ft}; L = 646.35 \text{ ft}; I = \frac{180L}{\pi R} = 35.95°$$

Tangent length, $T = R\tan\frac{I}{2} = 1{,}030 \times \tan\frac{35.95}{2} = 334.22 \text{ ft}$

Change in northing between PC and PI is calculated as:

$$\Delta N = T\cos Az = 334.22 \times \cos 222.5 = -246.41 \text{ ft}$$

Change in easting between PC and PI is calculated as:

$$\Delta E = T\sin Az = 334.22 \times \sin 222.5 = -225.80 \text{ ft}$$

Therefore, coordinates (north and east positive) of the PI are:

Northing $= 4{,}123.64 - 246.41 = 3{,}877.23$

Easting $= -1{,}064.32 - 225.80 = -1{,}290.12$

This can be expressed as $(3{,}877.23 \text{ N}, 1{,}290.12 \text{ W})$ (A)

Note: Since all answers have unique answer choices for both northings and eastings, calculating just one of them is adequate to identify the correct answer.

128

Point on curve (station $12 + 00$) can have maximum elevation $= 470.00 - 14.5 = 455.5$ ft

Since the PVI is completely specified (station as well as elevation), determine the coordinates of this point on the curve with respect to PVI:

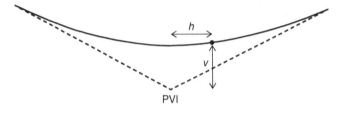

Horizontal offset $\qquad h = 12.0 - 10.563 = 1.437$ stations;

Vertical offset $\qquad v = 455.5 - 432.65 = 22.85$ ft

The following equation is useful when the PVI is used as the reference point:

$$\frac{L + 2h}{L - 2h} = \sqrt{\frac{v - G_1 h}{v - G_2 h}}$$

$$\frac{L + 2 \times 1.437}{L - 2 \times 1.437} = \sqrt{\frac{22.85 - (-4) \times 1.437}{22.85 - (+6) \times 1.437}} = \sqrt{\frac{28.598}{14.228}} = 1.41774$$

Solving: $L = 16.634$ sta. $= 1663.4$ ft (A)

Note: This is the MINIMUM length of curve. A shorter curve will not provide the necessary vertical clearance. So, if the answer choices were (A) 1,660, (B) 1,700, (C) 1,740, (D) 1,780, the answer would be 1,700 ft in spite of choice (A) 1,660 ft being closest to the solved value of 1,663.4 ft.

129

$$DHV = K \times AADT = 0.11 \times 78,500 = 8,635$$

Directional design hourly volume $V = DDHV = 0.57 \times 8,635 = 4,922$ vph

Approximate peak flow rate: $v_p = \dfrac{V}{PHF \times N} = \dfrac{4922}{0.94 \times 3} = 1,745 \dfrac{\text{veh}}{\text{hr}} /\text{ln}$ (B)

130

From the particle size distribution curve: $F_{200} = 12\%$

Since this is significant, AASHTO's equivalent to the Casagrande Plasticity Chart is used with LL = 34 & PI = 34 − 19 = 15

Table 203.5   AASHTO Soil Classification Criteria

Sieve Analysis	Granular Materials (35% or less passing no. 200 sieve)							Silt-Clay Materials (more than 35% passing no. 200 sieve)				
	A-1		A-3	A-2				A-4	A-5	A-6	A-7	A-8
	A-1-a	A-1-b		A-2-4	A-2-5	A-2-6	A-2-7					
% passing												
No. 10	≤50											
No. 40	≤30	≤50	>50									
No. 200	≤15	≤25	≤10	≤35	≤35	≤35	≤35	>35	>35	>35	>35	
LL				≤40	>40	≤40	>40	≤40	>40	≤40	>40	
PI		≤6	NP	≤10	≤10	>10	>10	≤10	≤10	>10	>10	
General description	Stone, gravel, sand		Fine sand	Silty or clayey gravel and sand				Silty soils		Clayey soils		Peat, highly organic soils
Quality as subgrade material	Good to excellent							Fair to poor				Very poor

Soil classification is A2-6 ($F_{200} < 35\%$)   (B)

In this question, answer choices do not include the group index, making it unnecessary. However, if there had been two choices of group A2-6, but with two different values of group index, the following would be necessary.

Group Index: $GI = 0.01(F_{200} - 15)(PI - 10) = -0.15$. This should be reported as non-negative. Therefore, $GI = 0$. The calculation of the group index is *shown but not necessary* for this problem. Also, note that the single-term expression above for the group index is recommended for groups A2-6 and A2-7. For other groups, the expression for the group index has an additional term.

Soil is classified as A2-6 (0).

131

$F_{200} = 10$. Since this is less than 50, it is predominantly a coarse-grained soil (first letter G or S). Also, since F_{200} is between 5 and 12%, soil will have dual classification. This eliminates (A)

Coarse fraction = 90%

$R_4 = 100 - 41 = 59$. This is more than half of the coarse fraction. Therefore, first letter is G.

Note, in these problems, solution strategy may depend heavily on given answer choices. All given choices have first letter G. Therefore, the last step is redundant.

$D_{10} = $ No. 200 size = 0.075 mm

$D_{30} = $ No. 10 size = 2.0 mm (slightly less)

$D_{60} = 0.5$ in = 12.7 mm

$$C_u = \frac{D_{60}}{D_{10}} = \frac{12.5}{0.075} = 166.7 \qquad\qquad C_c = \frac{D_{30}^2}{D_{10}D_{60}} = \frac{2.0^2}{0.075 \times 12.5} = 4.3$$

Since F_{200} is between 5% and 12%, the soil has a dual classification. The first part of the classification is based on gradation. Since BOTH criteria ($C_u > 4$ and $1 < C_c < 3$) for GW are not met, the soil must be classified GP. This eliminates (D)

GP-GM if PI < 0.73(LL–20) OR PI < 4

GP-GC if PI > 0.73(LL–20) AND PI > 7

PI = 54 – 23 = 31

Value of PI on the A-line = 0.73(LL – 20) = 24.8. Soil plots above the A-line (clay)

Thus, the soil meets BOTH criteria for GP-GC (B)

132

In this solution, the elevation datum is taken at the impenetrable layer. The total head at the upstream ground surface = 40 ft (gage pressure head) + 50 ft (elevation head) = 90 ft. Total head at downstream ground surface (under 5 ft of water) = 5 + 50 = 55 ft. Head loss through seepage $= 90 - 55 = 35$ ft. The total head at intermediate location X can be expressed as the weighted average (10.5 pressure drops out of 13) shown below:

$$TH_X = \frac{10.5}{13} \times 55 + \frac{2.5}{13} \times 90 = 61.73 \text{ ft}$$

Alternatively, one can also calculate the total head at X as the total head at upstream ground minus the head loss for 10.5 pressure drops.

$$TH_X = 90 - 35 \times \frac{10.5}{13} = 61.73 \text{ ft}$$

At point X, the elevation head is $EH_X = 12$ ft

Therefore, the pressure head is $PH_X = TH_X - EH_X = 61.73 - 12 = 49.73$ ft

This is equivalent to a pressure: $p_X = 49.73 \times 62.4 = 3,103 \text{ psf} = 21.55 \text{ psi}$ (D)

133

Statement I is correct. All other factors remaining equal, air entrainment reduces strength.

Statement II is correct. Air entrainment allows the concrete to expand/contract during freeze-thaw cycles.

Statement III is incorrect. If the concrete is not expected to experience alternating cycles of freeze-thaw, air entrainment is **not necessary.**

Statement IV is correct. A maximum aggregate size increases, mortar content decreases, thereby decreasing required air content.

Answer is   (D)

134

For $f'_c > 5,000$ psi, the required compressive strength f'_{cr} must be the greater of

$$f'_{cr} = f'_c + 1.34 s_s = 6,000 + 1.34 \times 675 = 6,905 \text{ psi and}$$

$$f'_{cr} = 0.9 f'_c + 2.33 s_s = 0.9 \times 6,000 + 2.33 \times 675 = 6,973 \text{ psi}$$

Therefore, f'_{cr} must be greater than 6,973 psi   (C)

135

GPR, nuclear density test, and the cone penetrometer test are all soil tests. Liquid penetrant test is typically used to detect surface defects of plastics, metals, and ceramics. The Brinell hardness test is used to measure the surface hardness of metals. Therefore, III and IV are not used for soils. (A)

136

Length of zone = 2,531.20 − 505.25 = 2,025.95 ft

Width of zone = 24 ft

Area subject to clearing and grubbing = 48,622.8 ft²

Cost = $9,724.56

Answer is (A)

137

The abbreviation CMP usually refers to Corrugated Metal Pipe. Reinforced Concrete Pipe is designated RCP. The only choice that doesn't fit the markings is (C)

138

Using trapezoidal method, the volume of excavation is calculated using:

$$V = \frac{\Delta}{2}\left[y_0 + y_n + 2\sum_1^{n-1} y_i \right]$$

$$V = \frac{50}{2} \times [456.33 + 493.34 + 2 \times (563.97 + 702.24 + 1234.98 + 783.92 + 591.94)]$$
$$= 217,594.25 \text{ ft}^3 = 8059 \text{ yd}^3 \qquad (B)$$

139

Using a 2V:1H plane of influence from the bottom edge of the footing (depth = 6.5 ft) to the bottom edge of the trench (depth = 12.0 ft) the 5.5 ft difference in elevation corresponds to a 2.75 ft horizontal distance. Since the right edge of the footing is at 3 ft, the edge of the trench can be no closer than 3 + 2.75 = 5.75 ft   (C)

140

According to the guidelines of the Standard Practice for Bracing Masonry Walls During Construction, the restricted zone should extend a width = H + 4 ft on either side of the wall.   (D)

Answer Key for Breadth Exam No. 2

101	D	111	D	121	A	131	B
102	D	112	B	122	B	132	D
103	B	113	B	123	B	133	D
104	A	114	C	124	D	134	C
105	C	115	A	125	C	135	A
106	D	116	C	126	D	136	A
107	A	117	B	127	A	137	C
108	C	118	A	128	A	138	B
109	A	119	D	129	B	139	C
110	C	120	C	130	B	140	D

10

Structural Depth Exam Solutions

These detailed solutions are for questions 201 to 240, representative of a 4-hr Structural Depth exam according to the syllabus and guidelines for the Principles and Practice (P&P) of Civil Engineering Examination administered by the National Council of Examiners for Engineering and Surveying (NCEES), current for the October 2020 examination.

201

For cantilever columns which are not allowed to rotate at the 'free' end, the load

displacement relationship is $\Delta = \dfrac{PL^3}{12EI}$.

(Continued on next page)

The stiffness is, therefore, $k = \dfrac{12EI}{L^3}$

The effective lumped weight is: $\dfrac{1}{2}M_{\text{column}} + M_{\text{floor}} = 0.5 \times 1.8 + 48 = 48.9$ k

Natural frequency:

$$\omega_n = \sqrt{\frac{k}{m}} = \sqrt{\frac{kg}{W}} = \sqrt{\frac{12EI}{L^3}\frac{g}{W}} = \sqrt{\frac{12 \times 29000 \times 10000 \times (32.2 \times 12)}{(14 \times 12)^3 \times 48.9}} = 76.2 \frac{\text{rad}}{\text{sec}}$$

Fundamental period: $T_n = \dfrac{2\pi}{\omega_n} = 0.083$ sec

Correct answer is (A)

202

Distributed load $w = 200 + 650 = 850$ lb/ft

Maximum shear force $V = wL/2 = 850 \times 24/2 = 10{,}200$ lb

Maximum shear stress $\tau = \dfrac{3V}{2bh} = \dfrac{3 \times 10200}{2 \times 5.5 \times 11.25} = 247.2$ psi

Answer is (B)

203

Maximum bending moment: $M_a = \dfrac{wL^2}{8} = \dfrac{6 \times 24^2}{8} = 432$ k · ft

Cracking moment: $M_{cr} = 115$ k · ft

Therefore, $(M_{cr}/M_a)^3 = 0.266^3 = 0.019$

Gross moment of inertia: $I_g = \dfrac{1}{12} \times 20 \times 28^3 = 36{,}587$ in^4

Effective moment of inertia: $I_e = \left[1 - \left(\dfrac{M_{cr}}{M_3}\right)^3\right]I_{cr} + \left(\dfrac{M_{cr}}{M_3}\right)^3 I_g = 0.981 \times 11{,}480 +$

$0.019 \times 36{,}587 = 11{,}960$ in^4

Modulus of elasticity of concrete: $E_c = 33\gamma_c^{1.5}\sqrt{f_c'} = 33 \times 150^{1.5} \times \sqrt{5,000} =$
4.3×10^6 psi

Maximum deflection (under load of $w = 500$ lb/in): $\Delta_{max} = \dfrac{5wL^4}{384EI} =$
$\dfrac{5 \times 500 \times 288^4}{384 \times 4.3 \times 10^6 \times 11960} = 0.87$ in

Answer is (A)

204

The plastic moment capacity of the beam: $M_P = Z_x F_y = 189 \times 50 = 9450$ k · in $=$
787.5 k · ft

At the time of collapse, plastic hinges must form at the fixed support and under the load P.

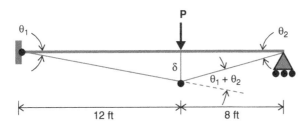

From compatibility of deformations: $\delta = 12\theta_1 = 8\theta_2 \Rightarrow \theta_2 = 1.5\theta_1$

Using the Principle of virtual work (equating external work to internal work)

$M_P\theta_1 + M_P(\theta_1 + \theta_2) = P\delta$

$M_P\theta_1 + M_P(\theta_1 + 1.5\theta_1) = P \times 12\theta_1 \Rightarrow 3.5M_P\theta_1 = 12P\theta_1 \Rightarrow P = 0.2917M_P$

Maximum load $P = 0.2917 \times 787.5 = 229.7$ k

Answer is (D)

205

Bending moment: $M_{max} = \dfrac{wL^2}{8} = \dfrac{275 \times 20^2}{8} = 13{,}750 \text{ lb} \cdot \text{ft} = 165{,}000 \text{ lb} \cdot \text{in}$

For a sawn 6×10 timber beam, section modulus $S_x = 82.729 \text{ in}^3$ (dressed dimensions $5.5 \times 9.5 \text{ in}$)

Bending stress: $\sigma = \dfrac{M}{S} = \dfrac{165000}{82.279} = 1{,}995 \text{ psi}$ (D)

206

The dressed dimensions of the column cross section (2×6 nominal) are $1.5 \text{ in} \times 5.5 \text{ in}$

For buckling about the x-axis, $K = 2.10$ (recommended—NDS)

Slenderness ratio: $\dfrac{KL}{r} = \dfrac{2.1 \times 12 \times 12}{0.29 \times 5.5} = 189.6$

For buckling about the y-axis, $K = 0.80$ (recommended—NDS)

Slenderness ratio: $\dfrac{KL}{r} = \dfrac{0.8 \times 12 \times 12}{0.29 \times 1.5} = 264.8$

Euler buckling load, $P_E = \dfrac{\pi^2 EA}{\left(\dfrac{KL}{r}\right)^2} = \dfrac{\pi^2 \times 1.5 \times 10^6 \times 1.5 \times 5.5}{264.8^2} = 1{,}741 \text{ lb} = 1.74 \text{ k}$

The stress level is 211 psi (for most grades of lumber, this ensures elastic behavior)

Correct answer is (A)

207

The vertical reaction at A can be calculated by taking moments about G:

$A_y = \dfrac{16 \times 75 + 26 \times 60 + 30 \times 45 + 26 \times 30 + 16 \times 15}{90} = 57 \text{ k}$

However, note that in this case, the structure *and* the load are symmetric; therefore, the two reactions will be equal to half the total vertical load of 114 k. Also note that if the end-game (of taking moments about A) is seen ahead of time, then it is obvious that the vertical reaction at A will not be necessary and therefore one can avoid that calculation.

Taking moments about A (so that the intersecting forces F_{CD} and F_{KJ} are not involved):

$$\frac{4}{5}F_{CI} \times 30 + \frac{3}{5}F_{CI} \times 20 + 16 \times 15 + 26 \times 30 = 0 \Rightarrow F_{CI} = -28.3\,\text{k}$$

Correct answer is (D)

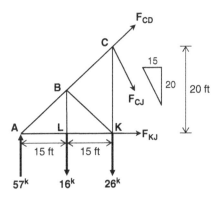

208

Since gravity forces counteract uplift, we must consider the empty tank case here. The weight of 6 k is carried equally by each leg. Therefore, the compression at each footing = 1.5 k.

The lateral force creates an overturning moment = 120 k × 65 ft = 7,800 k-ft on the horizontal plane at the top of the footings. This overturning moment is shared equally by two couples (four legs in two pairs). Therefore, each resisting couple = 3,900 k-ft.

Since the lever arm on each couple is 30 ft, the force at each footing = 3900 ÷ 30 = 130 k. This is an added compression under the legs on the far side (right) of the tower and an uplift under the legs on the near side (left).

Therefore, the near side legs experience a 'net' uplift of 130 − 1.5 = 128.5 k (B)

209

Taking moments about D:

$$\sum M_D = 8 \times 20 - 18B_y + 36 \times 4.5 - 12 \times 6 = 0$$

$$B_y = \frac{8 \times 20 + 36 \times 4.5 - 12 \times 6}{18} = 13.89$$

Using equilibrium of vertical forces, $D_y = 8 + 36 + 12 - 13.89 = 42.11$

The free body diagram is followed by the shear force and bending moment diagrams. The areas under the shear diagram are used to generate the bending moment diagram. For each segment, the order (constant, linear, quadratic etc.) of the moment diagram is *one higher* than that of the shear diagram.

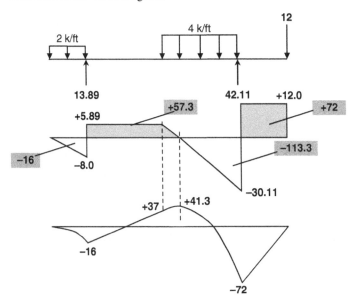

The maximum bending moment is 72 k-ft (B)

210

Solve for truss member forces using equilibrium of joint B:

$$\sum F_x = 0 \Rightarrow F_{AB} = F_{CB}$$

$$\sum F_y = 0 \Rightarrow 2 \times \frac{10}{\sqrt{164}} \times F_{AB} = 30 \text{ k} \Rightarrow F_{AB} = +19.21 \text{ k (tension)}$$

$$F_{AB} = F_{BC} = +19.21 \text{ k (T)}$$

Create the Virtual Load (consistent with desired deflection).

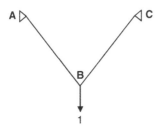

Calculate the member forces due to virtual load (*Note:* in this case, the real load and the virtual load are congruent, so we can use scaling)

$$f_{AB} = f_{BC} = +19.21 \text{ k} \div 30 \text{ k} = +0.64 \text{ (T)}$$

According to the Principle of Virtual Work, the vertical deflection at B is given by:

$$\Delta = \sum \frac{FfL}{AE} = \frac{19.21 \times 0.64 \times (12.81 \times 12)}{2 \times 29000} + \frac{19.21 \times 0.64 \times (12.81 \times 12)}{3 \times 29000} = 0.054 \text{ in}$$

Correct answer is (A)

Note: The positive sign of the answer indicates that this deflection is in the same sense as the assumed unit load (downward).

211

Knowledge of the nodal moments M_{AB}, M_{BA}, M_{BC} etc. allows the "decoupling" of the individual spans, as shown below.

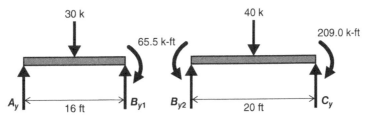

(*Continued on next page*)

For the left substructure, taking moments about A yields B_{y1}:

$$M_A = 30 \times 8 + 65.5 - 16B_{y1} = 0 \Rightarrow B_{y1} = 19.094\,k$$

For the right substructure, taking moments about C yields B_{y2}:

$$M_C = 209 - 65.5 - 40 \times 10 + 65.5 + 20B_{y2} = 0 \Rightarrow B_{y2} = 12.825\,k$$

Therefore, total vertical reaction at $B = B_{y1} + B_{y2} = 19.094 + 12.825 = 31.919\,k$ \hfill (D)

212

Using the unit load method, the deflection at a point is given by the integral:

$$\delta = \int \frac{Mm}{EI} dx$$

where M is the bending moment function under actual load and m is the bending moment function under virtual load. These are shown in the following figures.

Flexural rigidity $EI = 29000 \times 870 = 2.523 \times 10^7\,\text{k-in}^2 = 1.75 \times 10^5\,\text{k-ft}^2$

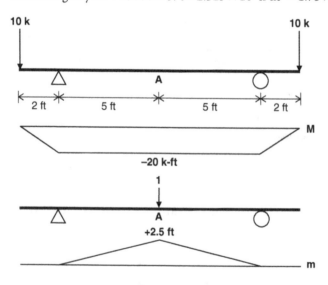

$$\delta = 2 \times \left[\frac{1}{2} \times \frac{-20 \times 2.5 \times 5}{EI} \right] = -\frac{250}{EI} = -\frac{250}{1.75 \times 10^5} = -0.001429\,\text{ft} = -0.017\,\text{in}$$

In the step above, the integral is facilitated by the use of Table 102.1 in the *All-in-One* book.

The negative sign indicates that the deflection is opposite to the direction of the unit load (which was applied downward) (C)

213

The propped cantilever has a degree of indeterminacy = 1. Therefore, for a plastic hinge collapse mechanism to form, *two* plastic hinges must form. These will occur at the locations of maximum moment, that is, at the support and at the point where the point load P acts. At this condition, the collapse mechanism will be as shown in the following figure. Assuming rotations θ_1 and θ_2 at A and C, the compatibility equation for the deflection at B is: $12\theta_1 = 3\theta_2$

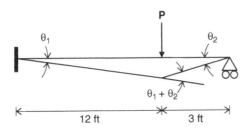

According to the Principle of Virtual Work, work done by internal moments acting through deformations = work done by external loads acting through beam displacements:

$$M_p\theta_1 + M_p(\theta_1 + \theta_2) = P\delta$$

$$M_p\theta_1 + M_p(\theta_1 + 4\theta_1) = P \cdot 12\theta_1 \Rightarrow M_p = 2P$$

For W18 × 90 with $F_y = 50$ ksi, plastic moment capacity, $M_p = Z_x F_y = 186 \times 50 = 9,300$ k-in = 775 k-ft

$$P = M_p/2 = 387.5 \text{ k}$$ (C)

214

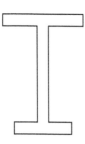

Depth of web = 22 − 1.5 − 2.0 = 18.5 in.

The total section area, $A = 10 \times 2 + 0.75 \times 18.5 + 12 \times 1.5 = 51.875$ in.^{2}

Half area = 25.94 in.^{2}

Area of bottom flange = 20 in^2. Therefore, web area below PNA = 25.94 − 20 = 5.94 in.^{2}

Depth of web below PNA = 5.94/0.75 = 7.917 in. This means the PNA is at a distance = 9.917 in above bottom edge.

The plastic section modulus Z_x is calculated as the first moment of the four component rectangles about the PNA:

$Z_x = 20 \times 8.917 + 5.94 \times 3.958 + 7.94 \times 5.292 + 18 \times 11.333 = 447.86$ in.^{3}

Plastic moment capacity: $M_p = Z_x F_y = 447.86 \times 36 = 16{,}123$ k · in = 1,343.6 k · ft   **(A)**

215

For f_c' greater than 5,000 psi, reduction of all content indicated in Table 4.4.1 by 1.0 percent.

Based on ACI Manual Table 4.4.1 (Air content) and Table 4.2.1 (Exposure condition)

Air content percentage = 6 − 1 = 5%   **(D)**

216

Initial prestress (at release) = $0.75 f_{pu}$ = 202.5 ksi

Prestressing force (after losses), $P_s = (202.5 − 32) \times 5.9 = 1{,}006$ k

Eccentricity of the prestress force = 21.65 in. The moment due to prestress force induces tension on the top fiber, while the moment due to gravity loads induces compression on top.

$M_{DL+LL} = 1400$ k-ft = 16,800 k-in

Stress on top fiber (compression positive) is calculated as:

$$\sigma_{top} = +\frac{P_s}{A} - \frac{P_s e}{S_t} + \frac{M}{S_t} = +\frac{1,006}{283.8} - \frac{1,006 \times 21.65}{2560} + \frac{16,800}{2560} = +1.6 \text{ ksi (compression)} \quad \text{(B)}$$

217

The allowable compressive stress of unreinforced masonry units subject to a combination of axial and bending stress is given by $0.45 f'_m$

For $f'_m = 1,800$ psi, the allowable compressive stress = 810 psi

Answer is   (D)

218

According to the IBC wind load provisions, a building is considered flexible if the fundamental frequency is less than 1 Hz. For the given building, the fundamental period is 0.35 sec. Therefore, the fundamental frequency is 1/0.35 = 2.86 Hz. This means, the building cannot be considered flexible.

According to the IBC wind load provisions, a building is considered low-rise if the mean roof height (in this case 42 ft) is less than 60 ft *and* least lateral dimension (which in this case is 80 ft). Therefore, this building can be considered low-rise.

Answer is   (D)

219

According to section 7.2.2 of ACI 318, minimum bend diameter of stirrups and ties no. 5 and smaller is $4d_b$. Since no. 4 bars have diameter $d_b = 0.5$ in, minimum bend diameter = 2.0 in. Therefore, minimum bend radius = 1.0 in.

Answer is   (A)

220

Maximum permitted reinforcement (ACI) = 8%

Factored load, $P_u = 1.2 \times 300 + 1.6 \times 350 = 920\text{ k}$

$$P_u \leq \phi \beta A_g [0.85 f_c'(1 - \rho_g) + \rho_g f_y]$$

$0.7 \times 0.85 \times A_g \times [0.85 \times 4 \times (1 - 0.08) + 0.08 \times 60] = 4.717 A_g \geq 920 \Rightarrow$
$A_g \geq 195\text{ in}^2$

Diameter ≥ 15.8 in. Choose D = 16 in   (C)

221

The HL-93 loading consists of: lane load of 640 lb/ft per loaded lane (10 ft wide) + (HS20) truck load in a design lane.

Therefore, per design lane, the maximum shear on a simple span is given by:

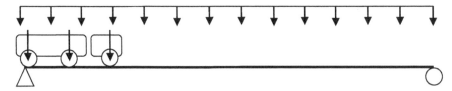

HL-93 loading = (HS25 tuck or Tandem Load whichever governs) + Lane Load

For the maximum shear at either support A or B, the distance between 2nd and 3rd wheels (which varies) should be minimum, that is, 14 ft.

Lane Load = 0.64 k/ft

Then, maximum shear = vertical reaction at the left support (for the load placement shown) = contribution of truck wheels (left to right) + contribution of lane load:

$$V_{max} = 32 + \frac{125 - 14}{125} \times 32 + \frac{125 - 14 - 14}{125} \times 8 + \frac{0.64 \times 125}{2} = 106.62\text{ k} \qquad \text{(C)}$$

For tandem (two 25-k axles separated by 4 ft), the maximum shear is given by:

$$V_{max} = 25 + \frac{125 - 4}{125} \times 25 + \frac{0.64 \times 125}{2} = 89.2\text{ k}$$

222

Since the compression flange of the beam is braced at midspan, $L_{unbraced} = 30/2 = 15$ ft

Note: Use unbraced length to calculate the Maximum Factored Moment Capacity.

LRFD

From the AISC manual, for $L_b = 15$ ft, for the W16 × 100 section, $\phi_b M_n = 670$ k-ft

Max. factored moment capacity = 670 k-ft (approx.) (Using AISC manual Table 3-10)

For a simple span, maximum moment for uniformly distributed load:

$$M_u = \frac{w_u L^2}{8} \Rightarrow w_u = \frac{670 \times 8}{30^2} = 5.95 \frac{k}{ft}$$

ASD

From the AISC manual, for $L_b = 15$ ft, for the W16 × 100 section, $M_n/\Omega = 450$ k-ft

Max. moment capacity = 450 k-ft (approx.) (Using AISC manual Table 3-10)

For a simple span, maximum moment for uniformly distributed load:

$$M_a = \frac{w L^2}{8} \Rightarrow w = \frac{450 \times 8}{30^2} = 4.0 \frac{k}{ft}$$

Therefore, the correct answer is (B)

223

Factored axial load: $P_u = 1.2 P_D + 1.6 P_L = 1.2 \times 300 + 1.6 \times 180 = 648$ k

Factored moment, $M_u = 1.2 P_D e_D + 1.6 P_L e_L = 1.2 \times 300 \times 0 + 1.6 \times 180 \times 4 = 1{,}152$ k · in

The dimensionless parameters:

$$K_n = \frac{P_u}{\phi_c f'_c A_g} = \frac{648}{0.65 \times 4 \times 320} = 0.78$$

$$R_n = \frac{M_u}{\phi_c f'_c A_g h} = \frac{1{,}152}{0.65 \times 4 \times 320 \times 20} = 0.07$$

(Continued on next page)

Assuming effective cover of about 2.5 in, $\gamma h = 20 - 2 \times 2.5 = 15$ in $\rightarrow \gamma = 0.75$

For the above data ($f'_c = 4$ ksi, $f_y = 60$ ksi, $\gamma = 0.75$), the column interaction diagram yields $\rho_g = 1.2\%$.

Therefore, $A_s = 0.012 \times 16 \times 20 = 3.84$ in^2 **(B)**

224

ASD Solution

Service loads are as shown in the figure.

The vertical reaction at A can be calculated by taking moments about E:

$$A_y = \frac{120 \times 36 + 200 \times 24 + 120 \times 12 - 240 \times 9}{48} = 175 \text{ k}$$

Under these loads, $A_x = -240$ k, $A_y = 175$ k, $E_y = 265$ k

By beam analogy, the maximum tension in the bottom chord will occur at midspan (members BC and CD)

Using method of sections, $F_{BC} = 546.7$ k

Yield criterion: $0.6 F_y A_g \geq 546.7 \Rightarrow A_g \geq \dfrac{546.7}{0.6 \times 36} = 25.3$ in^2

Fracture criterion: $0.5 F_u A_e = 0.5 F_u (0.75 A_g) \geq 546.7 \Rightarrow A_g \geq \dfrac{546.7}{0.5 \times 0.75 \times 58} = 25.1$ in^2

LRFD Solution

Factored loads are as follows: at B and D: $1.2 \times 0.3 \times 120 + 1.6 \times 0.7 \times 120 = 177.6$ k

                                at C: $1.2 \times 0.3 \times 200 + 1.6 \times 0.7 \times 200 = 296$ k

                                at F: $1.2 \times 0.3 \times 240 + 1.6 \times 0.7 \times 240 = 355.2$ k

The vertical reaction at A can be calculated by taking moments about E:

$$A_y = \frac{177.6 \times 36 + 296 \times 24 + 177.6 \times 12 - 355.2 \times 9}{48} = 259 \text{ k}$$

Under these loads, $A_x = -355.2$ k, $A_y = 259$ k, $E_y = 392.2$ k

By beam analogy, the maximum tension in the bottom chord will occur at midspan (members BC and CD)

Using method of sections, $F_{BC} = 809.1$ k

Yield criterion: $0.9 F_y A_g \geq 809.1 \Rightarrow A_g \geq \dfrac{809.1}{0.9 \times 36} = 24.97 \text{ in}^2$

Fracture criterion: $0.75 F_u A_e = 0.75 F_u (0.75 A_g) \geq 809.1 \Rightarrow A_g \geq \dfrac{809.1}{0.75^2 \times 58} = 24.8 \text{ in}^2$

Correct answer is (D)

225

The factored load is $w_u = 1.2\, w_{DL} + 1.6\, w_{LL} = 8.8 \text{ k/ft}$

The design moment is $M_u = w_u L^2/8 = 687.5$ k-ft

We are looking for the *smallest* size of the beam, therefore for *maximum allowed reinforcement*. The strength reduction factor is being assumed to be 0.81 (maximum steel $\rightarrow$ minimum ϕ).

ACI318-05 gives $\phi = 0.48 + 83\varepsilon_t \leq 0.9$. Minimum permissible tensile strain $\varepsilon_t = 0.004 \rightarrow \phi = 0.812$.

Using the design table (Table 105.5) in the *All-in-One* book:

For $f_c' = 4$ ksi, $f_y = 60$ ksi, $\rho_{max} = 0.0206$, $w_{max} = \dfrac{0.0206 \times 60}{4} = 0.309 \rightarrow X = 0.2527$

Thus, the effective depth: $d = \sqrt{\dfrac{M_u}{0.81 X b f_c'}} = \sqrt{\dfrac{687.5 \times 12}{0.81 \times 0.2527 \times 15 \times 4}} = 25.9$ in

Assuming $d \approx h - 2.5$, minimum overall depth $= 25.9 + 2.5 = 28.4$ in (C)

226

a) Flange ratio

$\dfrac{b_f}{2 t_f} = \dfrac{12}{2 \times 2} = 3$

Compact limit, $\lambda_p = 0.38 \sqrt{\dfrac{E}{F_y}} = 0.38 \times \sqrt{\dfrac{29{,}000}{50}} = 9.15$ (AISC Table B4.1)

(*Continued on next page*)

Since $\dfrac{b}{t} = 3 < \lambda_p$, therefore, the flange is compact.

b) Web

$$\frac{h}{t_w} = \frac{48}{0.5} = 96$$

Compact limit: $\lambda_p = 3.76\sqrt{\dfrac{E}{F_y}} = 3.76 \times \sqrt{\dfrac{29{,}000}{50}} = 90.31$

Noncompact limit: $\lambda_r = 5.70\sqrt{\dfrac{E}{F_y}} = 5.70 \times \sqrt{\dfrac{29{,}000}{50}} = 137.27$

Since $\lambda_p < \dfrac{h}{t_w} < \lambda_r$, web is noncompact.

Since the web is noncompact, the whole section is noncompact.   (C)

227

The effective width of flange = smallest of ($L/4$, center to center spacing, $b_w + 12t$) = min (7 ft, 8 ft, 88 in) = 7 ft = 84 in

Cross section of beam & slab = $96 \times 5 + 28 \times 15 = 900$ in^2 = 6.25 ft^2

Self weight of beam & slab = 6.25 ft$^2 \times 0.15$ k/ft^3 = 0.94 k/ft

Floor live load transmitted to each beam = 85 psf $\times$ 8 ft = 680 lb/ft = 0.68 k/ft

Total factored load on beam = $1.2 \times 0.94 + 1.6 \times 0.68 = 2.216$ k/ft

Maximum moment: $M_u = \dfrac{wL^2}{8} = \dfrac{2.216 \times 28^2}{8} = 217.17$ k · ft = 2,606 k · in

For a rectangular beam with ($b = 84$ in, effective depth $d = 17.5$ in, $M_u = 2{,}606$ k-in, $f_c' = 4$ ksi and $f_y = 60$ ksi

Strength parameter: $X = \dfrac{M_u}{\phi f_c' bd^2} = \dfrac{2{,}606}{0.9 \times 4 \times 84 \times 17.5^2} = 0.0281$

From the table (Table 105.5 in *All-in-One* book), corresponding value of the reinforcement parameter: $w = \rho \dfrac{f_y}{f_c'} = 0.0285$

$$\rho = w \dfrac{f_c'}{f_y} = \dfrac{0.0285 \times 4}{60} = 0.0019$$

This is less than the minimum steel ratio, so according to ACI, provide 33% more than required steel.

Area of steel, $A_s = 1.33 \times 0.0019 \times 84 \times 17.5 = 3.72 \text{ in}^2$ **(B)**

228

ASD Solution

Required available strength is calculated as:

$P_a = 200 + 200 = 400 \text{ k}$

The W12 × 96 section has the following properties: $r_x = 5.44$; $r_y = 3.09$

Slenderness ratios are calculated as: $(KL/r)_x = 88.2$; $(KL/r)_y = 77.7$

From Table 4-22 of the AISC steel manual, for the governing slenderness ratio = 88.2, allowable stress is given by

$F_{cr}/\Omega_c = 14.28 \text{ ksi}$

Allowable compressive load is given by

$P_n/\Omega_c = 14.28 \times 28.2 = 402.7 \text{ k}$

LRFD Solution

$P_u = 1.2 \times 200 + 1.6 \times 200 = 560 \text{ k}$

The W12 × 96 section has the following properties: $r_x = 5.44$; $r_y = 3.09$

Slenderness ratios are calculated as: $(KL/r)_x = 88.2$; $(KL/r)_y = 77.7$

(Continued on next page)

From Table 4-22 of the AISC steel manual, for the governing slenderness ratio = 88.2, design buckling stress is given by

$\phi_c F_{cr} = 21.56 \text{ ksi}$

Design strength in compression is given by

$\phi_c P_n = 21.56 \times 28.2 = 608 \text{ k (OK)}$

Check the next lower size. W12 × 87: $r_x = 5.38$; $r_y = 3.07$; $(KL/r)_x = 89.2$; $(KL/r)_y = 78.2$; $\phi_c F_{cr} = 21.36 \text{ ksi}$; $\phi_c P_n = 21.36 \times 25.6 = 547 \text{ k (inadequate)}$

Choose W12 × 96 (C)

229

The effective span of the lintel is calculated as: S = center to center distance between bearings = 5.67 ft

Height of wall above the lintel = 8 ft > 5.67 ft. Therefore, we can assume that arching action occurs.

Thus, the lintel needs to carry two kinds of loading:

1. self-weight (rectangular): $w_1 = 140 \times \dfrac{7 \cdot 5}{12} \times \dfrac{16}{12} = 116.7 \text{ lb/ft}$

2. wall weight (triangular): $w_2 = 130 \times \dfrac{7 \cdot 5}{12} \times \dfrac{5.67}{2} = 230.3 \text{ lb/ft}$

Maximum bending moment in the simply supported lintel is given by:

$$M_{max} = M_1 + M_2 = \frac{w_1 L^2}{8} + \frac{w_2 L^2}{12} = 469.0 + 617.0 = 1{,}086 \text{ lb} \cdot \text{ft}$$

Correct answer is (D)

230

According to AASHTO, effective width of slab = center to center distance between beams = 8 ft = 96 in.

Modular ratio: $n = \dfrac{E_s}{E_c} = \dfrac{29{,}000 \text{ ksi}}{1{,}820\sqrt{4} \text{ ksi}} = 7.97 = 8$

Equivalent width of slab $= 96/8 = 12$ in

Height of centroid (measured from bottom of steel): $\bar{y} = \dfrac{136 \times 30 + 12 \times 8 \times 64}{136 + 96} =$ 44.1 in

Moment of inertia of equivalent section:

$$I_{NA} = 81{,}940 + 136 \times (44.1 - 30)^2 + \frac{1}{12} \times 12 \times 8^3 + 96 \times (44.1 - 64)^2 = 147{,}507 \text{ in}^4$$

Maximum bending moment in composite section: $M = \dfrac{wL^2}{8} = \dfrac{1 \cdot 8 \times 8 \times 70^2}{8} =$ 8,820 k · ft = 105,840 k · in

Bending stress in steel (tensile): $\sigma = \dfrac{My}{I} = \dfrac{105{,}840 \times 44 \cdot 1}{147{,}507} = 31.6$ ksi (A)

231

A572 grade 50 steel: $F_y = 50$ ksi; $F_u = 65$ ksi

For the channel (C10 × 30), relevant properties: $A_g = 8.81$ in^2; $t_w = 0.673$ in

Net area (2 holes): $A_{net} = 8.81 - 2 \times \dfrac{7}{8} \times 0.673 = 7.63$ in^2

Net area (3 holes): $A_{net} = 8.81 - 3 \times \dfrac{7}{8} \times 0.673 + 2 \times \dfrac{2^2}{4 \times 2 \cdot 5} \times 0.673 = 7.58$ in^2

Shear lag factor: $U = 1 - \dfrac{\bar{x}}{L} = 1 - \dfrac{0.649}{4} = 0.84$

ASD Solution

Capacity (yield): $\dfrac{P_n}{\Omega_t} = \dfrac{F_y A_g}{1.67} = \dfrac{50 \times 8.81}{1.67} = 263.8$ k

Capacity (fracture): $\dfrac{P_n}{\Omega_t} = \dfrac{F_u U A_{net}}{2.0} = \dfrac{65 \times 0.84 \times 7.58}{2.0} = 206.9$ k

Design member capacity = 207 k

(*Continued on next page*)

LRFD Solution

Capacity (yield): $\phi_t P_n = 0.9 F_y A_g = 0.9 \times 50 \times 8.81 = 396.5\text{ k}$

Capacity (fracture): $\phi_t P_n = 0.75 F_u A_e = 0.75 F_u U A_{net} = 0.75 \times 65 \times 0.84 \times 7.58 = 310.4\text{ k}$

Design member capacity $= 310\text{ k}$ (A)

232

Eccentricity is greater than 0.1 h (4 in > 0.1 × 16 in). Therefore, the column must be designed for a combination of P_u and M_u ($e/h = 4/16 = 0.25$)

Area of steel $A_s = 12 \times 1.0 = 12.0\text{ in}^2$, and gross area $A_g = 16 \times 16 = 256\text{ in}^2$

Reinforcement ratio $\rho_g = 12 \div 256 = 0.047$ (This is within limits 1%–8%)

Assuming clear cover = 1.5 in, the center to center distance between parallel lines of reinforcement $= 16 - 2 \times (1.5 + 0.5 + 1.128/2) = 10.87\text{ in}$

Parameter $\gamma = 10.87/16 = 0.68$. Let us use the diagram for $f'_c = 4$ ksi, $f_y = 60$ ksi, $\gamma = 0.65$

From the chosen column interaction diagram, for $e/h = 0.25$ and $\rho_g = 4.7\%$, $K_n = 0.80$

$$P_u = K_n \phi_c f'_c A_g = 0.80 \times 0.65 \times 4 \times 256 = 532.5\text{ k} \qquad \text{(D)}$$

233

Slab self-weight: $w_{sw} = 150 \times \dfrac{5}{12} = 62.5\text{ lb/ft}^2$

Total factored load on slab: $w_u = 1.2 \times (40 + 62.5) + 1.6 \times 85 = 259\text{ lb/ft}^2$

For a unit width (1 ft) of the slab, the factored moment at the positive moment critical section is:

$$M_u = \frac{w_u L^2}{10} = \frac{259 \times 8^2}{10} = 1{,}657.6\text{ lb}\cdot\text{ft} = 1{,}9891.2\text{ lb}\cdot\text{in}$$

Assuming #5 main reinforcement bars and 0.75-in clear cover, we get

Effective depth of slab $= 5.0 - 0.75 - 0.625/2 = 3.94\text{ in}$

If ultimate moment is expressed as: $M_u = \phi f_c' bd^2 X$, then the strength parameter X is given by

$$X = \frac{M_u}{\phi f_c' bd^2} = \frac{19891.2}{0.9 \times 4{,}500 \times 12 \times 3.94^2} = 0.0264 \Rightarrow w = \frac{\rho f_y}{f_c'} = 0.027$$

Reinforcement ratio, $\rho = \dfrac{w f_c'}{f_y} = 0.002$

Since this reinforcement is less than minimum required reinforcement, it must be increased by 33%

Required area of steel, $A_s = \rho bd = 1.33 \times 0.002 \times 12 \times 3.94 = 0.126 \text{ in}^2/\text{ft}$ **(C)**

234

In order for block shear rupture to occur, four blocks must rupture (two in each flange) in order for the connection to fail. One of these blocks is shown in the following figure.

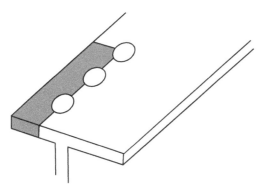

The block parameters (areas in tension and shear) are calculated as:

$$A_{nt} = \left(3 - \frac{1}{2} \times \frac{7}{8}\right) \times 0.67 = 1.717 \text{ in.}^2$$

$$A_{gv} = 12 \times 0.67 = 8.04 \text{ in.}^2$$

$$A_{nv} = \left(12 - \frac{5}{2} \times \frac{7}{8}\right) \times 0.67 = 6.57 \text{ in.}^2$$

$$R_n = 0.6 F_u A_{nv} + U_{bs} F_u A_{nt} \le 0.6 F_y A_{gv} + U_{bs} F_u A_{nt}$$

$R_n = 328.5 \le 273.3$. Thus, the governing nominal strength of one block $= 273.3$ k. This is the nominal strength of one (of four) blocks. Thus, nominal strength in block shear $= 4 \times 273.3 = 1093.2$ k **(B)**

235

According to ASCE 7-08

The correct answer choice is   (A)

236

The correct answer choice is   (C)

237

According to Table 2.4 of the AISC Manual of Steel Construction, the preferred ASTM grade for square HSS shapes is A500 Gr. C

Answer is   (A)

238

According to Figure D-11 in §1910.27, the optimal range of the pitch angle is between 60° and 75° to the horizontal.   (C)

239

The total dead load of the bridge deck + girders = $36 \times 80 \times 0.75 \times 150 + 4 \times 80 \times 282 = 324,000 + 90,240 = 414,240$ lb. This force acts at midspan.

Taking moments about the support far from the jacked end, total force on jacks = $414 \, k \times 40/70 = 237 \, k$

Assuming this load is equally shared by the six jacks, each jack needs to exert a force of $39.5 \, k = 19.7$ tons.

If a minimum $FS = 2.0$ is desired, the jacks must be rated at 39.5 tons.

Answer is   (B)

240

At the rate of return (i), the present worth should be zero.

The \$350k capital expenditure is a present value (P) NEGATIVE

The \$25k reduction in annual costs is an annuity (A) POSITIVE

The \$200k increase in salvage value is a future sum (F) POSITIVE

Converting all of these to present worth, the net present worth can be written:

$$PW = -350 + 25\left(\frac{P}{A}, i, 20 \text{ yrs}\right) + 200\left(\frac{P}{F}, i, 20 \text{ yrs}\right) = 0$$

For $i = 5\%$, $PW = 37$k

For $i = 6\%$, $PW = -0.9$k   Actual answer 5.97%

For $i = 7\%$, $PW = -33.5$k

Answer is   (B)

Answer Key for Structural Depth Exam

201	A		211	D		221	C		231	A
202	B		212	C		222	B		232	D
203	A		213	C		223	B		233	C
204	D		214	A		224	D		234	B
205	D		215	D		225	C		235	A
206	A		216	B		226	C		236	C
207	D		217	D		227	B		237	A
208	B		218	D		228	C		238	C
209	B		219	A		229	D		239	B
210	A		220	C		230	A		240	B

11

Geotechnical Depth Exam Solutions

These detailed solutions are for questions 301 to 340, representative of a 4-hr Geotechnical Depth exam according to the syllabus and guidelines for the Principles and Practice (P&P) of Civil Engineering Examination administered by the National Council of Examiners for Engineering and Surveying (NCEES), current for the October 2020 examination.

301

Weight of test sand in test hole = 13.75 − 10.24 = 3.51 lb

Volume of test sand in test hole = 3.51 ÷ 88.2 = 0.0398 ft^3

Weight of soil obtained from the test hole = 5.52 lb

Unit weight of soil obtained from the test hole = 5.52 ÷ 0.0398 = 138.71 lb/ft^3

Dry unit weight of soil = 138.71 ÷ 1.19 = 116.56 lb/ft^3

Percent compaction = 116.56 ÷ 126.3 = 92% (A)

302

The cumulative % retained on the no. 200 sieve is 74.

$F_{200} = 100 - R_{200} = 100 - 74 = 26$

Since $F_{200} < 50$, soil is predominantly coarse. Therefore, first letter is S or G.

Coarse fraction = $100 - F_{200} = 74$. Half the coarse fraction = 37

$R_4 = 8$ is less than half the coarse fraction. Therefore, first letter is S.

Since $F_{200} > 12$, second letter of classification should be based on plasticity characteristics rather than gradation. Use the Casagrande Plasticity chart with LL = 43, PI = 43 − 21 = 22. This lies above the A-line (more like clay than silt).

Soil is classified as SC.   (D)

303

The sum of lengths of all unfractured pieces longer than 4 in = 214 in

Total core length = 20 ft = 240 in

RQD = 214 ÷ 240 = 89%   (D)

304

Statement B is true. Seismic reflection methods are useful for identifying deep features, but are of limited use in shallow ground investigation.

Statement C is true. Gravity surveys are conducted by measuring minute variations in the earth's gravitational force. This requires very delicate and expensive equipment.

Statement D is true. Resistivity surveys are of limited use in shallow ground investigation because of the difficulty in interpreting the results.

Answer is   (A)

305

The friction ratio = 0.03/5.2 = 0.6%. Such a low value usually indicates sand. **(A)**

306

Weight of wet soil = 1,331.5

The weight of the wax = 1,368.2 − 1,331.5 = 36.7 g. Using the specific gravity of the wax, we can calculate the wax volume = 36.7 ÷ 0.9 = 40.78 cm^3

Since the water content = 15.2%, the weight of the soil can be split into dry soil solids = 100/115.2 × 1,331.5 = 1,155.8 g and water (175.7 g).

Since the SG of soil solids is known (2.70), we also have volume of soil solids = 1,155.8 ÷ 2.7 = 428.07 cm^3 and volume of water = 175.7 cm^3.

Also, the buoyancy (apparent loss of weight on immersion) of the wax-coated sample = 1,368.2 − 593.4 = 774.8 g. Therefore, the volume of displaced water = volume of wax-coated sample = 774.8 cm^3 (Density of water = 1 g/cm^3).

Therefore, the volume of the (uncoated) soil sample = 774.8 − 40.78 = 734.02 cm^3

Therefore, volume of air = 734.02 − (428.07 + 175.7) = 130.25 cm^3

Total volume of voids in the original sample = 175.7 + 130.25 = 305.95 cm^3

Degree of saturation $S = V_{water}/V_{voids}$ = 175.7 ÷ 305.95 = 0.574 **(A)**

307

With the porous end-plates, pore pressures do not build up. Effective stress is equal to total stress. Area of plane experiencing shear failure = $\pi(2)^2$ = 12.56 in^2

Completing the table with the normal ($\sigma = N/A$) and shear ($\tau = F/A$) stresses, we have:

Sample	N (lb)	F (lb)	σ (psi)	τ (psi)
1	120	78	9.55	6.21
2	160	93	12.73	7.40
3	220	116	17.51	9.23

Using sample points 1 and 3, the cohesion is calculated as:

$$c = \frac{\sigma_2\tau_1 - \sigma_1\tau_2}{\sigma_2 - \sigma_1} = \frac{17.51 \times 6.21 - 9.55 \times 9.23}{17.51 - 9.55} = 2.587 \text{ psi} = 372.5 \text{ psf}$$ **(A)**

308

The depth associated with this SPT result is at the center of the 2nd and 3rd penetration intervals, that is, exactly 1 ft below the start depth for the test. Thus, the depth is $10 + 1 = 11$ ft.

The effective vertical stress at $z = 11$ ft is:

$$\sigma'_v = 109.2 \times 6 + (122.7 - 62.4) \times 2 + (118 - 62.4) \times 3 = 942.6 \text{ psf} = 0.471 \text{ tsf}$$

Overburden correction factor (Liao & Whitman): $C_N = \sqrt{\dfrac{1}{\sigma'_v}} = \sqrt{\dfrac{1}{0.471}} = 1.46$

Corrected SPT N value: $N' = C_N N = 1.46 \times 39 = 57$ (C)

309

The hydraulic gradient in the sand drain is calculated from the head difference and the total length:

$$i = \frac{H}{L} = \frac{945 - 905}{160} = 0.25 \text{ ft/ft}$$

Sand porosity: $n = \dfrac{e}{1 + e} = \dfrac{0.45}{1 + 0.45} = 0.31$

Seepage velocity (this is the true fluid velocity through the voids in the soil, rather than an effective velocity across the entire cross section) is given by:

$$v_s = \frac{Ki}{n} = \frac{1 \times 10^{-4} \times 0.25}{0.31} = 8.1 \times 10^{-5} \text{ ft/sec}$$

Scour velocity $= 8$ in/hr $= 1.85 \times 10^{-4}$ ft/sec

Factor of safety against scour is given by:

$$FS = \frac{1.85 \times 10^{-4}}{8.1 \times 10^{-5}} = 2.29 \qquad \text{(C)}$$

310

Sample 1 (undrained)—effective stress analysis

Total axial stress at failure is calculated as a sum of the radial chamber pressure and the added load of 158.2 lb.

$$\sigma_1 = 18 + \frac{158.2}{\frac{\pi}{4}(2)^2} = 68.36 \text{ psi}$$

Effective vertical stress is given by

$$\sigma_1' = 68.36 - 5.6 = 62.76 \text{ psi}$$

Effective radial stress is given by

$$\sigma_3' = 18 - 5.6 = 12.4 \text{ psi}$$

Assuming angle of internal friction $= 0$, cohesion: $c = \dfrac{\sigma_1' - \sigma_3'}{2} = \dfrac{62.76 - 12.4}{2} = 25.18 \text{ psi}$

Sample 2 (drained, pore pressure zero)—total stress analysis

Radial stress, $\sigma_3' = 36 \text{ psi}$

$$c = \frac{\sigma_1' - \sigma_3'}{2} \Rightarrow \sigma_1' = \sigma_3' + 2c = 36 + 2 \times 25.18 = 86.36 \text{ psi}$$

Since there is no pore pressure, this is also the value of the total vertical stress. Therefore, the axial deviatoric stress $= 86.36 - 36 = 50.36 \text{ psi}$

Corresponding axial load at failure: $P_f = 50.36 \times \dfrac{\pi}{4} \times 2^2 = 158.2 \text{ lb}$ (D)

311

By examination, it seems that the peak dry unit weight will come from sample 3 or 4.

Volume of Standard Proctor mold $= 1/30 \text{ ft}^3$

For sample 3, total unit weight, $\gamma = W/V = 3.95/(1/30) = 118.5 \text{ lb/ft}^3$

Dry unit weight, $\gamma_d = 118.5/1.16 = 102.2 \text{ lb/ft}^3$

(Continued on next page)

For sample 4, total unit weight, $\gamma = W/V = 4.21/(1/30) = 126.3 \text{ lb/ft}^3$

Dry unit weight, $\gamma_d = 126.3/1.18 = 107.0 \text{ lb/ft}^3$ (maximum dry unit weight)

Target dry unit weight $= 0.9 \times 107 = 96.3 \text{ lb/ft}^3$

Weight of solids needed $= 96.3 \times 1.5 \times 10^6 \times 27 = 3.9 \times 10^9 \text{ lb}$

From table, for 12% moisture content, unit weight $= 3.24/(1/30) = 97.2 \text{ lb/ft}^3$; $\gamma_d = 97.2/1.12 = 86.8 \text{ lb/ft}^3$.

Volume of borrow soil needed $=$
$3.9 \times 10^9 \text{ lb} \div 86.8 \text{ lb/ft}^3 = 4.49 \times 10^7 \text{ ft}^3 = 1.66 \times 10^6 \text{ yd}^3$ **(B)**

312

The compacted soil in the embankment has dry density $0.95 \times 95 \text{ pcf} = 90.25 \text{ pcf}$.

500,000 yd^3 of embankment volume contains solids $= 500{,}000 \times 27 \times 90.25 = 1.218 \times 10^9 \text{ lb}$.

The borrow soil (while being transported) has a total density of 115 pcf and a water content of 20%. Therefore, the dry density during transport $= 115/1.2 = 95.83 \text{ pcf}$.

Total weight of soil during transport $= 1.20 \times 1.218 \times 10^9 = 1.462 \times 10^9 \text{ lb}$

Therefore, the volume of the soil (in transport) $= 1.218 \times 10^9 \text{ lb} \div 95.83 = 1.271 \times 10^7 \text{ ft}^3$

Truck carries $V = 10 \text{ yd}^3 = 270 \text{ ft}^3$, which weighs $270 \times 115 = 31{,}050 \text{ lb}$

Number of trips based on weight $= 1.462 \times 10^9 \text{ lb} \div 31{,}050 \text{ lb} = 47{,}072$

Number of trips based on volume $= 1.271 \times 10^7 \text{ ft}^3 \div 270 \text{ ft}^3 = 47{,}072$ **(B)**

313

The Rankine failure plane is oriented at $\alpha = 45 + \phi/2$ as shown in the figure. From the back face of the wall, the distance $X = 5 + 16/\tan 62 = 13.5 \text{ ft}$. Any load on the backfill must be placed beyond this surface.

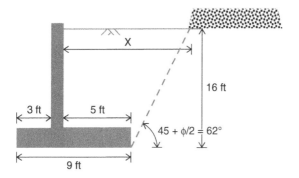

Answer is   (C)

314

At a depth of 30 ft, effective vertical stress is given by:

$$\sigma'_v = 124 \times 10 + (124 - 62.4) \times 20 = 2{,}472 \text{ psf}$$

Average cyclic shear stress: $\tau_{ave} = \text{CSR} \times \sigma'_v = 0.23 \times 2{,}472 = 568.6 \text{ psf}$

Ultimate shear strength of the soil in the field (corrected for relative density) is:

$$\tau_{ult} = \frac{0.85}{0.95} \times 1{,}200 = 1{,}074 \text{ psf}$$

Factor of safety for liquefaction is given by:

$$FS = \frac{\tau_{ult}}{\tau_{ave}} = \frac{1074}{568.6} = 1.89 \qquad\qquad (B)$$

315

The weighted average of the SPT N values for the upper 100 ft is calculated as:

$$\bar{N} = \frac{100}{\sum \dfrac{d_i}{N_i}} = \frac{100}{\dfrac{12}{12} + \dfrac{20}{34} + \dfrac{33}{25} + \dfrac{23}{33} + \dfrac{12}{22}} = 24.1$$

Based on seismic site characterization (ASCE7 criteria) is D ($15 < N < 50$)   (D)

316

1926 Subpart P, App. A: Soil with unconfined compression strength greater than 1.5 tons/ft² (3,000 lb/ft²) is classified as type A unless it is subject to vibration (and some other restrictive conditions) in which case it is classified as type B.

Answer is   (D)

317

Using the Taylor Stability Chart (shown in the following figure), using $\beta = 30$, $D = 50/20 = 2.5$, Stability number, $N_0 = 5.6$

Factor of safety (slope stability): $FS = \dfrac{N_0 c}{\gamma H} = \dfrac{5.6 \times 900}{120 \times 20} = 2.1$ (C)

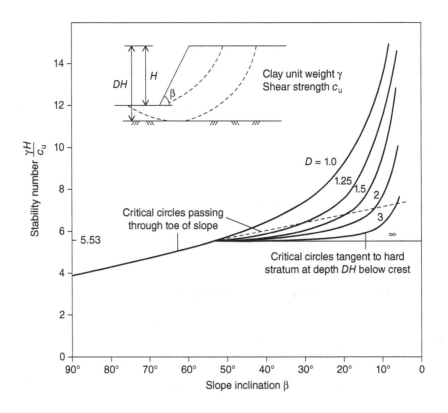

318

The structural number is given by (excluding any effects of drainage):

$$SN = a_1 D_1 + a_2 D_2 + a_3 D_3 = 0.45 \times 4 + 0.25 \times 8 + 0.15 \times 12 = 5.6 \qquad \text{(B)}$$

319

The clay layer can be considered to be "singly drained" since the underlying clay layer can be considered impermeable. Therefore, the drainage thickness H_d = the full layer thickness = 12 ft

For degree of consolidation $U = 80\%$, time factor $T_v = 0.567$

The time for 80% settlement to occur is given by

$$t = \frac{T_v H_d^2}{c_v} = \frac{0.567 \times 12^2}{6} = 13.6 \text{ yrs} \qquad \text{(C)}$$

320

Discharge $Q = 2{,}000$, gpm = 4.464 ft^3/sec

Hydraulic conductivity $K = 1{,}000$ ft/day $= 1.157 \times 10^{-2}$ ft/sec

At observation well 1, radial distance $r_1 = 30$ ft, height of piezometric surface (above aquifer bottom): $y_1 = 50 - 4.5 = 45.5$ ft

At observation well 2, radial distance $r_2 = 180$ ft, height of piezometric surface (above aquifer bottom): $y_2 = ?$

The steady-state equation for the drawdown due to pumping in an aquifer is:

$$Q = \frac{\pi K \left(y_1^2 - y_2^2 \right)}{\ln \left(\dfrac{r_1}{r_2} \right)}$$

where y_1 and y_2 are water table elevations at radial (horizontal) distance r_1 and r_2 (from pump well centerline), respectively.

$$Q = \frac{\pi K \left(y_1^2 - y_2^2 \right)}{\ln \left(\dfrac{r_1}{r_2} \right)} \Rightarrow \left(y_1^2 - y_2^2 \right) = \frac{Q \ln \left(\dfrac{r_1}{r_2} \right)}{\pi K} = \frac{4.464 \times \ln \left(\dfrac{30}{180} \right)}{\pi \times 1.157 \times 10^{-2}} = -220$$

Solving, we get $y_2 = 47.9$ ft. Drawdown $s_2 = 50 - 47.9 = 2.1$ ft \qquad (A)

321

Head difference: $H = 213.45 - 198.65 = 14.80$ ft

Number of flow channels = 3; Number of potential drops = 8

By Darcy's law, seepage flow per unit width is given by:

$$q = K\frac{N_f}{N_e}H = 200 \times \frac{3}{8} \times 14.8 = 1{,}110 \ \frac{\text{ft}^2}{\text{day}} = 0.01285 \ \frac{\text{ft}^2}{\text{sec}}$$

Total seepage loss under dam: $Q = qL = 0.01285 \times 150 = 1.93 \ \dfrac{\text{ft}^3}{\text{sec}}$ (C)

322

The setup shows the falling head permeability test. The key aspect for this setup is the choice of datum. The elevation of the downstream end of the soil sample is 229.80 ft. This should be the reference elevation. Therefore, at $t = 0$, $h_1 = 238.56 - 229.80 = 8.76$ ft

$$K = \frac{aL\ln\left(\dfrac{h_1}{h_2}\right)}{A\Delta t} \Rightarrow \ln\left(\frac{h_1}{h_2}\right) = \frac{KA\Delta t}{aL} = \frac{0.012 \times \dfrac{\pi}{4} \times 2^2 \times \dfrac{10}{60}}{\dfrac{\pi}{4} \times 0.1^2 \times 1}$$

$$= 0.8 \Rightarrow \frac{h_1}{h_2} = 2.23 \Rightarrow h_2 = 3.94 \text{ ft}$$

Therefore, at $t = 10$ min, elevation of water column $= 229.80 + 3.94 = 233.74$ ft   (D)

323

I is not a true statement. Karst is produced from the dissolution of soluble rocks.

II is a true statement. Karst formations consist of dolomitic soils that lower pH of water that comes in contact with it.

III is not true. Karst formations are highly porous and do not provide adequate bearing.

IV is true. Because of the high porosity, Karst formations allow water to seep into subsurface reservoirs faster than other soils.

Answer is   (A)

324

Fine-grained soils, such as clays, have much lower electrical resistivity than sands. This leads to a high corrosion potential for cast iron pipes embedded in these soils.

Answer is (D)

325

Of the given parameters, the following are significant in determining the frost heave of soils—grain size (soils with no particles smaller than 74 microns do not heave), pore size, presence of moisture combined with freeze-thaw cycles caused by temperature fluctuations and overburden pressure. (C)

326

Consider a unit width (1 ft) of the sheet piling. The load per unit length = 500 psf × 1 ft = 500 plf

The sheet pile is considered to be hinged at the wales, and continuous over more than three wales, the design moment is given by:

$$M = \frac{wL^2}{10} = \frac{500 \times 4^2}{10} = 800 \text{ lb} \cdot \text{ft} = 9{,}600 \text{ lb} \cdot \text{in}$$

Required section modulus: $S = \dfrac{M}{\sigma_{all}} = \dfrac{9{,}600}{1{,}400} = 6.86 \text{ in}^3$

For a rectangular section: $S = \dfrac{bt^2}{6} = 6.86 \Rightarrow t = \sqrt{\dfrac{6.86 \times 6}{12}} = 1.85 \text{ in}$ (D)

327

For clays, the cohesion is half the unconfined compression strength: $c = S_{uc}/2 = 600$ psf

Stiffness parameter: $\gamma H/c = 115 \times 24/600 = 4.6 > 4.0$. Therefore, assume it to be soft clay. According to Peck, the lateral pressure in a braced cut in soft clay grows linearly for the upper $H/4$ of the trench depth and then stays constant at that maximum value, which is approximately given by $\gamma H - 4c = 115 \times 24 - 4 \times 600 = 360$ psf

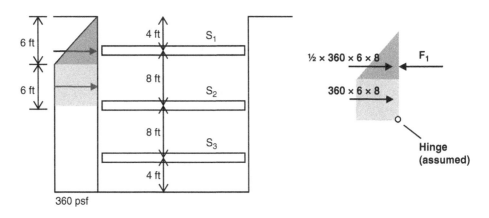

360 psf

Taking moments about S_2 (assume hinge at second strut location), the moment of the earth pressure resultants (shown shaded) about S_2 is balanced by the moment of the strut compression. Note that the earth pressure terms include the out-of-plane 8 ft strut spacing.

$$360 \times 6 \times 8 \times 3 + \frac{1}{2} \times 360 \times 6 \times 8 \times 8 = F_1 \times 8 \Rightarrow F_1 = 15{,}120 \text{ lb} \qquad \textbf{(B)}$$

328

Active earth pressure coefficient ($\beta = 0$, $\theta = 90$, $\delta = 0$, $\phi = 34$): $K_a = 0.283$

Active earth pressure resultant: $R_a = \frac{1}{2} K_a \gamma H^2 = 0.5 \times 0.283 \times 125 \times 15^2 = 3{,}980$ lb/ft

Overturning moment: $M_a = \frac{1}{6} K_a \gamma H^3 = \frac{1}{6} \times 0.283 \times 125 \times 15^3 = 19{,}898.4$ lb · ft/ft

Stabilizing moment is calculated as the sum of the moments of all weight components (wall footing, wall stem, and soil block above heel) about the toe:

$$M_s = \sum W_i x_i = 9 \times 3 \times 150 \times 4.5 + 12 \times 1 \times 150 \times 3.5 + 5 \times 12 \times 125 \times 6.5 = 73,275 \text{ lb} \cdot \text{ft/ft}$$

$$FS_{OT} = 73,275 \div 19,898.4 = 3.7 \tag{A}$$

329

All lateral and vertical force components are shown in the figure. A vertical plane is drawn through the heel of the footing.

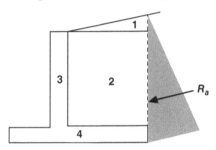

The height H, measured vertically from the bottom of the footing to the top of the backfill

$$H = 17 + 6 \tan 15 = 18.6 \text{ ft}$$

For $\phi = 34°$, $\beta = 15°$, $\delta = 0°$, $\theta = 90°$, Rankine's active earth pressure coefficient $K_a = 0.339$

Earth pressure resultant: $R_a = \dfrac{1}{2} K_a \gamma H^2 = 0.5 \times 0.339 \times 120 \times 18.6^2 = 7037 \dfrac{\text{lb}}{\text{ft}}$, which can be resolved into its vertical and horizontal components as follows:

$$R_{av} = 7,037 \times \sin 15 = 1,821 \text{ lb/ft}$$

$$R_{ah} = 7,037 \times \cos 15 = 6,797 \text{ lb/ft}$$

The weights of the soil wedges 1, 2, and the concrete elements 3 and 4 are calculated as follows, with their horizontal distances (measured from the toe):

$$W_1 = \frac{1}{2} \times 6 \times 1.6 \times 120 = 576 \text{ @ 9 ft}$$

$$W_2 = 14 \times 6 \times 120 = 10,080 \text{ @ 8 ft}$$

$$W_3 = 1 \times 14 \times 150 = 2,100 \text{ @ 4.5 ft}$$

$$W_4 = 11 \times 3 \times 150 = 4,950 \text{ @ 5.5 ft}$$

$$R_{av} = 1,821 \text{ @ 11 ft}$$

(Continued on next page)

The sum of the stabilizing moments about the toe is given by:

$$M_s = \sum F_i x_i = 576 \times 9 + 10{,}080 \times 8 + 2{,}100 \times 4.5 + 4{,}950 \times 5.5 + 1{,}821 \times 11$$
$$= 142{,}530 \text{ lb} \cdot \text{ft/ft}$$

The overturning moment about the toe is given by:

$$M_{OT} = R_{ah} \times \frac{H}{3} = 6797 \times \frac{18.6}{3} = 42{,}141 \text{ lb} \cdot \text{ft/ft}$$

$$FS_{OT} = 142{,}530/42{,}141 = 3.38 \tag{C}$$

330

$$K_a = \frac{1 - \sin\phi}{1 + \sin\phi} = 0.307$$

At depth of 6 ft, effective vertical pressure $= 123 \times 6 = 738$ psf

Effective horizontal pressure $= 0.307 \times 738 = 226.6$ psf

Total horizontal pressure $= 226.6 + 0 = 226.6$ psf

At depth of 35 ft, effective vertical pressure $= 123 \times 6 + (123 - 62.4) \times 29 = 2495.4$ psf

Effective horizontal pressure $= 0.307 \times 2{,}495.4 = 766.1$ psf

Total horizontal pressure $= 766.1 + 62.4 \times 29 = 2{,}575.7$ psf

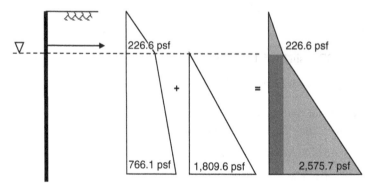

Therefore, the total horizontal earth pressure diagram grows from zero at the surface, to 226.6 psf at a depth of 6 ft, to 2,575.7 psf at a depth of 35 ft. This can be broken up into a triangle from $z = 0$ to $z = 6$, a rectangle from $z = 6$ to $z = 35$ and a triangle from $z = 6$ to $z = 35$. The resultants of these pressure diagrams are calculated (in order) as follows:

Resultant earth pressure = ½ × 226.6 × 6 + 226.6 × 29 + ½ × 29 × 2349.1 = 41,313 lb/ft (C)

331

Eccentricity: $e = M/P = 9/10 = 0.9$ ft. In order to determine whether the entire footing width is effective or not, the eccentricity must be compared to $B/6$, which is $4/6 = 0.67$ ft.

Since this eccentricity is greater than $B/6$, the resultant load does not fall within the middle third of the footing base. This will cause an uplift on the far side of the footing. Maximum soil pressure is given by:

$$q_{max} = \frac{4P}{3(B - 2e)} = \frac{4 \times 10}{3 \times (4 - 2 \times 0.90)} = 6.06 \text{ ksf} \qquad \text{(C)}$$

332

Depth of water table below footing = 5 ft, which is less than the footing width (10 ft). Therefore, it **will** reduce the bearing capacity.

Since the water table is below the bottom of the footing, the average unit weight to be used in the third term of the bearing capacity equation is given by:

$$\gamma_{ave} = \frac{\gamma D + (\gamma - \gamma_w)(B - D)}{B} = \frac{115 \times 5 + (115 - 62.4) \times (10 - 5)}{10} = 83.8 \text{ pcf}$$

Ultimate bearing capacity (assuming given bearing capacity factors include shape factors) is:

$$q_{ult} = cN_c + \gamma D_f N_q + \frac{1}{2}\gamma_{ave}BN_\gamma = 115 \times 2 \times 25 + 0.5 \times 83.8 \times 10 \times 19 = 13,711 \text{ psf}$$

Soil pressure due to column load + soil overburden = 750,000/100 + 115 × 2 = 7,730 psf

$FS = 13,711/7,730 = 1.77$ (B)

333

Footing size B = 5 ft. Normalizing with respect to footing size, we want the stress increase at 0.8B below and 1.2B lateral offset. This point is on the stress contour $0.05p = 0.05 \times 100/25 = 0.2 \text{ ksf} = 200 \text{ psf}$ (C)

334

Since the GWT is within 5 ft (footing width) of the bottom of the footing, it must be accounted for.

In the third term of the bearing capacity equation, replace γ with γ_{ave}

$$\gamma_{ave} = \frac{\gamma D + (\gamma - \gamma_w)(B - D)}{B} = \frac{125 \times 4 + (125 - 62.4) \times (5 - 4)}{5} = 112.5$$

Using Terzaghi's bearing capacity factors: For $\phi = 35°$, $N_c = 58$, $N_q = 42$, $N_\gamma = 46$

Ultimate bearing capacity for a square footing:

$$q_{ult} = 1.3cN_c + \gamma DN_q + 0.4\gamma BN_\gamma = 0 + 125 \times 3 \times 42 + 0.4 \times 112.5 \times 5 \times 46 = 26,100 \text{ lb/ft}^2$$

Allowable bearing pressure: $q_{all} = \dfrac{q_{ult}}{FS} = \dfrac{26,100}{2.8} = 9,321 \text{ lb/ft}^2$

Allowable load: $Q_{all} = q_{all}A_f = 9,321 \times 25 = 233,025 \text{ lb} = 233 \text{ k}$ (D)

335

Using the given dimensions, the following eccentricities of the column loads are as shown in the following figure:

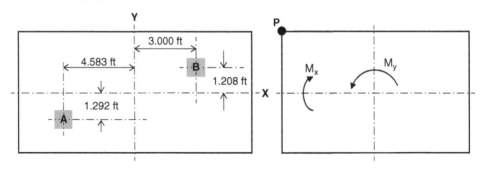

Resultant moment about the x-axis: $M_x = 120 \times 1.208 - 100 \times 1.292 = 15.76$ k-ft

Resultant moment about the y-axis: $M_y = 120 \times 3.000 - 100 \times 4.583 = -98.3$ k-ft

Under these moments, the maximum compression (under corner P) is given by

$$q = \frac{220}{13.167 \times 6.75} + \frac{15.76 \times 3.375}{\frac{1}{12} \times 13.167 \times 6.75^3} + \frac{98.3 \times 6.583}{\frac{1}{12} \times 6.75 \times 13.167^3}$$

$$= 2.475 + 0.158 + 0.504 = 3.137$$

Maximum vertical compression (incl. 4 ft of overburden pressure) =
$3{,}137 + 4 \times 127 = 3{,}645$ lb/ft^2 (B)

336

The dead load + wind load case produces maximum uplift.

Total dead load = 660 k, which produces compression in each pile = $660/27 = 24.4$ k

Wind load overturning moment is calculated as the sum of the moments of lateral forces about the top of the pile cap.

$$M_{OT} = 30 \times 15 + 60 \times 30 + 70 \times 45 + 80 \times 60 = 10{,}200 \text{ k} \cdot \text{ft}$$

Assuming linear force distribution in the columns (spacing S) and taking moments about pile line 5, we have 6 piles (lines 4 and 6) at distance S, 6 piles (lines 3 and 7) at distance $2S$, 6 piles (lines 2 and 8) at distance $3S$ and 6 piles (lines 1 and 9) at distance $4S$. The sum of these moments $= 6 \times F \times S + 6 \times 2F \times 2S + 6 \times 3F \times 3S + 6 \times 4F \times 4S = 180FS = 10{,}200$

$F = 10{,}200/(180 \times 6) = 9.44$ k. Thus, each pile in the outermost line of piles experiences a wind-uplift $= 4F = 37.78$ k

Gross uplift $= 37.78 - 24.44 = 13.34$ k

Net uplift = gross uplift − pile self-weight $= 13.34 - 5.6 = 7.74$ k (A)

337

Dimensions of pile group (between outside edges) = 22 ft × 10 ft

Center of clay layer is at depth = 37 ft

2/3 L of piles is at depth = 30.67 ft

Therefore, stress propagation through $\Delta z = 6.33$ ft

Bottom of clay layer at depth = 54 ft. Therefore, thickness of clay deposit undergoing consolidation settlement = 54 − 30.67 = 23.33 ft

Using a 2:1 stress propagation pyramid, where a loaded area $B \times L$ gets distributed over an area $(B + z)(L + z)$. Distribute pile group load over an area $(22 + 11.67)(10 + 11.67) =$ 33.67 ft × 21.67 ft

$$\Delta p = \frac{125 \times 2,000}{33.67 \times 21.67} = 342.6 \text{ psf}$$

Initial effective vertical stress: $p_1' = 120 \times 20 + 115 \times 17 = 4,355 \text{ lb/ft}^2$

After construction, effective vertical stress: $p_2' = 4,355 + 342.6 = 4,697.6 \text{ lb/ft}^2$

Consolidation settlement:

$$s = \frac{HC_c}{1 + e_0} \log_{10}\left(\frac{p_2'}{p_1'}\right) = \frac{23.33 \times 12 \times 0.4}{1.44} \log_{10}\left(\frac{4697.6}{4355}\right) = 2.56 \text{ in} \qquad \text{(B)}$$

338

Moment of inertia of pile group about group centerline (calculated as the second moment of all areas, $I = \Sigma A \cdot d^2$)

$$I = 4 \times A_p \times \left(\frac{B}{2}\right)^2 + 4 \times A_p \times \left(\frac{3B}{2}\right)^2 = 10 A_p B^2$$

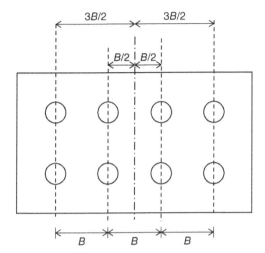

The maximum compressive force in outer piles (pair to the right):

$$P = A\sigma = A\left(\frac{P}{nA} + \frac{M \times 1.5B}{10AB^2}\right) = \frac{P}{n} + \frac{M \times 1.5}{10B} = \frac{350}{8} + \frac{0.15 \times 300}{B}$$

Equating this to 30 tons = 60 k, we get B_{min} = 2.77 ft. Therefore, B = 3 ft 0 in.   (B)

339

Design capacity = 40 tons

Since FS = 6, ultimate capacity = 40 × 6 = 240 tons = 480,000 lb

Pile hammer energy WH = 50,000 ft-lb = 600,000 in-lb

$$Q_{ult} = \frac{WH}{S + 1.0} \Rightarrow S = \frac{WH}{Q_{ult}} - 1 = \frac{600,000}{480,000} - 1 = 0.25$$

$$S = 0.25\frac{in}{blow} \Rightarrow 4\frac{blows}{in} = 48\frac{blows}{ft}$$ (D)

340

At the rate of return (i), the present worth should be zero.

The $350k capital expenditure is a present value (P)	NEGATIVE
The $25k reduction in annual costs is an annuity (A)	POSITIVE
The $200k increase in salvage value is a future sum (F)	POSITIVE

Converting all of these to present worth, the net present worth can be written:

$$PW = -350 + 25\left(\frac{P}{A}, i,\ 20\ \text{yrs}\right) + 200\left(\frac{P}{F}, i,\ 20\ \text{yrs}\right) = 0$$

For $i = 5\%$, $PW = 37k$

For $i = 6\%$, $PW = -0.9k$ \qquad Actual answer 5.97%

Answer is \hfill (B)

Answer Key for Geotechnical Depth Exam

301	A
302	D
303	D
304	A
305	A
306	A
307	A
308	C
309	C
310	D

311	B
312	B
313	C
314	B
315	D
316	D
317	C
318	B
319	C
320	A

321	C
322	D
323	A
324	D
325	C
326	D
327	B
328	A
329	C
330	C

331	C
332	B
333	C
334	D
335	B
336	A
337	B
338	B
339	D
340	B

12

Water Resources & Environmental Depth Exam Solutions

These detailed solutions are for questions 401 to 440, representative of a 4-hr Water Resources & Environmental Depth exam according to the syllabus and guidelines for the Principles and Practice (P&P) of Civil Engineering Examination administered by the National Council of Examiners for Engineering and Surveying (NCEES), current for the October 2020 examination.

401

If the bypass factor is x, then the fraction $1 - x$ gets treatment (and removal), while the fraction x gets no removal.

Incoming hardness quantity − Transmitted hardness = Removed hardness

$Q \times 200 - Q \times 50 = 0.88 \times Q \times (1 - x) \times 200 \Rightarrow x = 0.15$

Bypass fraction = 15%   (B)

402

The TSS in the primary effluent $= 0.4 \times 400 = 160$ mg/L

Assuming the primary sludge has specific gravity which is basically same as water, 5% solids (by weight) is equivalent to 50,000 mg/L.

Assuming that the primary sludge flow rate is Q_{ps} and performing a mass balance at the primary clarifier:

$2.5 \times 400 = (2.5 - Q_{ps}) \times 160 + 50,000 \times Q_{ps} \Rightarrow Q_{ps} = 0.012$

0.012 MGD $= 12,000$ gpd

Answer is (C)

403

Surface loading velocity $= 10$ ft/hr $= 0.002778$ ft/sec

Incident flow rate $Q = 3.2$ MGD $= 3.2 \times 1.5472$ ft^3/sec

Based on hydraulic load, total area of filter needed: $A = \dfrac{Q}{V} = \dfrac{3.2 \times 1.5472}{0.002778} = 1{,}782.4$ ft^2

Solid load in influent: $X = 800\dfrac{\text{mg}}{\text{L}} \times 3.2 \ MGD \times 8.3454 = 21{,}364\dfrac{\text{lb}}{\text{day}}$

Based on solid load, total area of filter needed: $A = \dfrac{21{,}364}{15} = 1{,}424.3$ ft^2. The other criterion governs.

Number of filters needed $= 1{,}782.4/250 = 7.13$ At least eight filters are needed. (C)

404

Orifice area, $A_o = \dfrac{\pi}{4}\left(\dfrac{2}{12}\right)^2 = 0.0218$ ft^2

Pipe area, $A_1 = \dfrac{\pi}{4}\left(\dfrac{4}{12}\right)^2 = 0.0873$ ft^2

Discharge coefficient of an orifice meter is given by:

$$C_f = \frac{C_v C_c}{\sqrt{1 - C_c^2 A_0^2 / A_1^2}} = \frac{0.95 \times 0.9}{\sqrt{1 - 0.9^2 \times 0.0218^2 / 0.0873^2}} = 0.878$$

$$Q = C_f A_o \sqrt{\frac{2g\Delta p}{\gamma}} = 0.878 \times 0.0218 \times \sqrt{\frac{2 \times 32.2 \times 30 \times 144}{62.4}} = 1.28 \text{ cfs} = 574 \text{ gal/min} \quad \textbf{(B)}$$

405

Extended Bernoulli's equation:

$$\frac{P_1}{\gamma} + \frac{V_1^2}{2g} + Z_1 - \Delta h = \frac{P_2}{\gamma} + \frac{V_2^2}{2g} + Z_2 \Rightarrow \Delta h = \frac{P_1 - P_2}{\gamma} + \frac{V_1^2 - V_2^2}{2g} + Z_1 - Z_2$$

Flow rate $Q = 650$ gpm $= 1.448$ cfs

Velocity upstream of reducer (8-in diameter) $V_1 = 4.15$ fps

Velocity downstream of reducer (4-in diameter) $V_2 = 16.60$ fps

Pressure loss: $P_1 - P_2 = 12$ psi $= 1,728$ psf, which is equivalent to a pressure head difference $= 27.69$ ft

$$\Delta h = 27.69 + \frac{4.15^2 - 16.6^2}{2 \times 32.2} + 0 = 23.7 \quad \textbf{(D)}$$

406

The head loss due to friction, expressed as a pressure loss (psi per foot) is given by:

$$\frac{\Delta P}{L} = \frac{4.52 Q^{1.85}}{C^{1.85} D^{4.87}} = \frac{4.52 \times 1,500^{1.85}}{110^{1.85} \times 12^{4.87}} = 0.00315 \quad \textbf{(B)}$$

where the flow rate Q (gpm) and diameter D (in)

407

Using stations 1 and 2 at the free surface of the reservoir and at the nozzle outlet respectively $(p_1 = p_2 = p_{atm})$ and ignoring head loss at the nozzle inlet:

$$z_1 + \frac{V_1^2}{2g} + \frac{p_1}{\gamma} = z_2 + \frac{V_2^2}{2g} + \frac{p_2}{\gamma} \Rightarrow V_2 = \sqrt{2g(z_1 - z_2)} = \sqrt{2 \times 32.2 \times \frac{40}{12}} = 14.65 \text{ fps}$$

Ideal flow rate: $Q_o = V_2 A_o = 14.65 \times \dfrac{\pi}{4} \times \left(\dfrac{0.5}{12}\right)^2 = 0.02$ cfs

Actual flow rate: $Q = C_d Q_o = 0.81 \times 0.02 = 0.016$ cfs $= 7.26$ gpm   **(A)**

408

Flow rate, $Q = 3,000$ gpm $= 3,000 \div 448.8 = 6.684$ ft³/s

For suction line:

Cross-section area, $A = \dfrac{\pi(1.5)^2}{4} = 1.767$ ft²; Velocity, $V = \dfrac{Q}{A} = \dfrac{6.684}{1.767} = 3.782$ ft/sec

Total head loss: $h_f + h_m = f\dfrac{L}{d}\dfrac{V^2}{2g} + \sum K\dfrac{V^2}{2g}$

$h_f + h_m = 0.024 \times \dfrac{800}{1.5} \times \dfrac{3.782^2}{2 \times 32.2} + 5 \times \dfrac{3.782^2}{2 \times 32.2} = 3.95'$

For discharge line:

Cross-section area, $A = \dfrac{\pi(1.0)^2}{4} = 0.785$ ft²; Velocity, $V = \dfrac{Q}{A} = \dfrac{6.684}{0.785} = 8.51$ ft/sec

Total head loss: $h_f + h_m = 0.026 \times \dfrac{2,500}{1.0} \times \dfrac{8.51^2}{2 \times 32.2} + 25 \times \dfrac{8.51^2}{2 \times 32.2} = 101.2'$

Total dynamic head $=$ static head $+$ head loss $= 80 + 4 + 101 = 185$ ft

Pump power rating: $P = \dfrac{\gamma Q H}{\eta} = \dfrac{62.4 \times 6.684 \times 185}{0.88} = 87,682$ lb · ft/sec $= 159.4$ hp   **(C)**

409

No matter how the flow distributes through the network, the head loss is proportional to Q^2 (if Darcy-Weisbach model is used) or $Q^{1.85}$ (if Hazen-Williams model is used). When the inflow changes from 300 gpm to 500 gpm, the (total) head loss becomes:

$$h_f = 70 \times \left(\frac{500}{300}\right)^2 = 194.4 \text{ ft} \qquad (180 \text{ ft if Hazen-Williams model is used})$$

Since 1 atm = 14.7 psi = 33.9 ft of water, 194.4 ft of head loss = 84.3 psi
(78.3 psi if Hazen-Williams model is used) (D)

410

This is an "unknown-depth-of-flow" problem. It is best solved using the tables (*All-in-One PE Exam Guide* Table 303.3 & equation 303.37).

Flow parameter: $K = \dfrac{Qn}{kb^{8/3}S^{1/2}} = \dfrac{150 \times 0.016}{1.486 \times 10^{8/3} \times 0.004^{1/2}} = 0.055$

For $m = 2$ (side slope parameter), the fragment of the table is shown here:

Table 303.3 Value of Parameter K for Straight-Sided Open Channels

d/b	0	0.25	0.5	0.75	1.0	1.5	2.0	2.5	3.0	3.5	4.0
					Horizontal projection m						
0.01	0.0005	0.0005	0.0005	0.0005	0.0005	0.0005	0.0005	0.0005	0.0005	0.0005	0.0005
0.02	0.0014	0.0014	0.0015	0.0015	0.0015	0.0015	0.0015	0.0015	0.0015	0.0015	0.0015
0.04	0.0044	0.0045	0.0046	0.0046	0.0047	0.0047	0.0048	0.0048	0.0049	0.0049	0.0050
0.06	0.0085	0.0087	0.0089	0.0090	0.0091	0.0093	0.0095	0.0096	0.0098	0.0099	0.0101
0.08	0.0135	0.0139	0.0142	0.0145	0.0147	0.0152	0.0155	0.0159	0.0162	0.0165	0.0168
0.1	0.0191	0.0198	0.0204	0.0209	0.0214	0.0221	0.0228	0.0234	0.0241	0.0247	0.0253
0.12	0.0253	0.0265	0.0275	0.0283	0.0290	0.0303	0.0314	0.0324	0.0334	0.0345	0.0355
0.14	0.0320	0.0338	0.0352	0.0365	0.0376	0.0395	0.0412	0.0428	0.0444	0.0459	0.0475
0.16	0.0392	0.0416	0.0437	0.0455	0.0471	0.0498	0.0523	0.0546	0.0569	0.0591	0.0614
0.18	0.0467	0.0500	0.0529	0.0553	0.0575	0.0612	0.0646	0.0678	0.0710	0.0741	0.0772
0.2	0.0547	0.0589	0.0627	0.0659	0.0687	0.0737	0.0783	0.0826	0.0868	0.0910	0.0952
0.22	0.0629	0.0683	0.0731	0.0772	0.0809	0.0874	0.0932	0.0989	0.1043	0.1098	0.1152
0.24	0.0714	0.0781	0.0841	0.0893	0.0939	0.1021	0.1096	0.1167	0.1237	0.1306	0.1374

Interpolating, $d/b = 0.164$

$d = 1.64 \text{ ft} = 19.7 \text{ in}$ (D)

411

$Q = 120,000$ gpm $= 268$ ft³/s

Flow per unit width $q = 268/15 = 17.87$ ft²/s

Critical depth: $d_c = \left(\dfrac{q^2}{g}\right)^{1/3} = \left(\dfrac{17.87^2}{32.2}\right)^{1/3} = 2.15$ ft

Therefore, the depth of 15 in (1.25 ft) is supercritical. A hydraulic jump occurs at this location as long as the tailwater depth d_2 corresponds to $d_1 = 1.25$.

The velocity at the bottom of the spillway is: $V_1 = \dfrac{Q}{bd} = \dfrac{268}{15 \times 1.25} = 14.3$ ft/sec

$$d_2 = -\frac{1}{2}d_1 + \sqrt{\frac{2V_1^2 d_1}{g} + \frac{d_1^2}{4}} = -\frac{1.25}{2} + \sqrt{\frac{2 \times 14.3^2 \times 1.25}{32.2} + \frac{1.25^2}{4}} = 3.41 \text{ ft} \qquad \textbf{(B)}$$

412

For pipe diameter $= 36$ in and $Q = 100$ cfs, FHWA nomograph yields $HW/D = 3.5$.

Headwater $= 3.5 \times 3 = 10.5$ ft (this is measured from the upstream invert elevation).

100-year water surface elevation (WSE) $= 325 + 10.5 = 335.5$ ft

Vertical clearance between roadway and 100-year WSE $= 337.8 - 335.5 = 2.3$ ft $\qquad$ **(B)**

413

Treating this as a broad-crested weir, the weir coefficient may be found (see table):

For crest breadth (measured parallel to flow) $= 12$ in $= 1$ ft and measured head $= 9$ in $= 0.75$ ft, the weir coefficient $C = 2.82$

Measured	Weir crest breadth w (ft)										
head (ft)	0.5	0.75	1.0	1.5	2.0	2.5	3.0	4.0	5.0	10.0	15.0
0.2	2.80	2.75	2.69	2.62	2.54	2.48	2.44	2.38	2.34	2.49	2.68
0.4	2.92	2.80	2.72	2.64	2.61	2.60	2.58	2.54	2.50	2.56	2.70
0.6	3.08	2.89	2.75	2.64	2.61	2.60	2.68	2.69	2.70	2.70	2.70
0.8	3.30	3.04	2.85	2.68	2.60	2.60	2.67	2.68	2.68	2.69	2.64
1.0	3.32	3.14	2.98	2.75	2.66	2.64	2.65	2.67	2.68	2.68	2.63
1.2	3.32	3.20	3.08	2.86	2.70	2.65	2.64	2.67	2.66	2.69	2.64
1.4	3.32	3.26	3.20	2.92	2.77	2.68	2.64	2.65	2.65	2.67	2.64
1.6	3.32	3.29	3.28	3.07	2.89	2.75	2.68	2.66	2.65	2.64	2.63

Flow rate: $Q = CbH^{3/2}$

Flow rate per unit width: $\dfrac{Q}{b} = CH^{\frac{3}{2}} = 2.82 \times 0.75^{3/2} = 1.83$ cfs/ft   (C)

414

For a rectangular open channel, critical velocity is given by: $V_c = \sqrt{gd_c} = 15$

Therefore, critical depth $d_c = 15^2 \div 32.2 = 6.99$ ft

Flow area $= 69.9$ ft^2. Flow rate, $Q = VA = 15 \times 69.9 = 1{,}048$ ft^3/s   (B)

415

During the 3 years of construction, the volume of sediment captured by the pond:
$V = 1{,}200 \times 280 \times 3 = 1{,}008{,}000$ ft^3

Depth of sediment in pond during first 3 years: $d = \dfrac{V}{A} = \dfrac{1.008 \times 10^6}{5 \times 43{,}560} = 4.63$ ft

Original depth of water in pond $= 353.4 - 345.6 = 7.8$ ft

Allowable depth of sediment (before sediment removal) $= 7.8 - 3.0 = 4.8$ ft.

Since 4.63 ft of sediment accumulates in first 3 years, remaining depth (before dredging) $= 0.17$ ft. This corresponds to a volume $= 0.17 \times 5 \times 43{,}560 = 37{,}026$ ft^3

Additional time to accumulate this volume of sediment $= t = \dfrac{37{,}026}{280 \times 300} = 0.44$ years

Number of years until sediment must be removed from pond $= 3.44$ years   (B)

416

Using the NRCS approach, we can calculate the runoff depth.

The composite curve number is given by:

$$\overline{CN} = \frac{\sum CN_i A_i}{\sum A_i} = \frac{69 \times 80 + 45 \times 80 + 98 \times 50 + 87 \times 90 + 35 \times 70}{370} = 66$$

For $CN = 66$ and a gross rainfall $P_g = 5.6$ in

Soil storage capacity, $S = \dfrac{1,000}{66} - 10 = 5.15$

Runoff depth, $Q = \dfrac{(P_g - 0.2S)^2}{P_g + 0.8S} = \dfrac{(5.6 - 0.2 \times 5.15)^2}{5.6 + 0.8 \times 5.15} = 2.15$ in **(D)**

The TR55 documentation also has a figure that can be used to look up runoff depth Q for a given value of gross rainfall P_g and curve number CN.

417

The simplest method for determining base flow is the "constant base flow" approach, which seems appropriate (since the discharge at $t = 0$ and $t = 6$ are nearly equal). After the base-flow ($Q_b = 23$) is subtracted, we get the pattern of net discharge caused by storm runoff:

Time (hr)	0	1	2	3	4	5	6
Discharge Q (ft³/sec)	0	61	104	89	52	9	2

Volume of runoff: $V = \Sigma Q \times \Delta t = 317 \times 3,600 = 1,141,200$ ft³

Average depth of runoff (excess precipitation) = $1,141,200/(115 \times 43,560) = 0.228$ ft = 2.73 in

The peak **runoff** discharge (ft³/sec) due to a 2-hr storm producing 1.7 in of runoff is most nearly: $104 \times 1.7/2.73 = 65$ ft³/sec

Adding in the base flow (unaffected by storm), the corresponding peak discharge recorded in stream = $65 + 23 = 88$ ft³/sec **(A)**

418

The following values of curve number are obtained from TR55 (Urban Hydrology for Small Watersheds).

Region	Area (acres)	Soil Type	Land Use	CN
1	80	C	Single family homes on ½ acre lots	80
2	50	D	Lawns in good condition	80
3	10	B	Paved streets and sidewalks	98
4	50	C	Grassy areas: fair condition	79
5	40	A	Woods: fair condition	36

Composite curve number:

$$\overline{CN} = \frac{\sum CN_i \times A_i}{\sum A_i} = \frac{80 \times 80 + 80 \times 50 + 98 \times 10 + 79 \times 50 + 36 \times 40}{230} = 72.9 \quad \text{(C)}$$

419

Time of concentration = time for sheet flow + time for channel flow = 200/0.5 + 400/2.0 = 600 s = 10 min

Using Steel formula (for Reno, NV, which is in region 7), for 50-year recurrence interval, coefficients are: $K = 65$, $b = 8$

Design rainfall intensity, $I = \dfrac{K}{t_c + b} = \dfrac{65}{10 + 8} = 3.61 \dfrac{\text{in}}{\text{hr}}$

For paved surfaces, use a rational coefficient = 0.98. Area = $400 \times 400 = 160{,}000$ ft^2 = 3.673 acres.

Runoff: $Q = CIA_d = 0.98 \times 3.61 \times 3.673 = 12.99 \dfrac{\text{acre} \cdot \text{in}}{\text{hr}} = 13.1$ ft^3/sec \quad \text{(A)}

420

Assuming average inflow during 5-min analysis period $= 50 \text{ ft}^3/\text{sec}$

During this stage, head at broad crested weir $= 126.4 - 125.0 = 1.4 \text{ ft}$ (*Note*: This is the weir head at the beginning of the analysis period and this rough solution assumes this stays constant, even though it does not. A better solution would be to break down the analysis period into smaller intervals.)

Effective width of (rectangular) weir, $b_e = b - 0.1NH = 5 - 0.1 \times 2 \times 1.4 = 4.72 \text{ ft}$

Assuming weir coefficient $= 3.3$, average outflow during this stage:
$Q = CbH^{3/2} = 3.3 \times 4.72 \times 1.4^{1.5} = 25.8 \text{ cfs}$

Net inflow during analysis period (5 min) $= 50 - 25.8 = 24.2 \text{ cfs}$

Net storage volume increase into pond during this period (5 min) $= 24.2 \times 5 \times 60 = 7,260 \text{ ft}^3$

Depth increase $= 7,260/43,560 = 0.167 \text{ ft}$

Pool elevation $= 126.4 + 0.167 = 126.567 \text{ ft}$

Weir head $= 126.567 - 125.0 = 1.567 \text{ ft}$

$Q = CbH^{3/2} = 3.3 \times 4.72 \times 1.567^{1.5} = 30.6 \text{ cfs}$ (A)

421

Since the end contractions are not suppressed, the effective width of the weir opening is:

$b_{\text{eff}} = b - 0.1nH = 6 - 0.1 \times 2 \times 5.4 = 4.92 \text{ ft}$

Weir discharge is calculated from:

$Q = 3.33 b_{\text{eff}} H^{3/2} = 3.33 \times 4.92 \times 5.4^{3/2} = 205.6 \dfrac{\text{ft}^3}{\text{sec}}$ (D)

422

I is false because above the water table, the total unit weight and not the buoyant unit weight should be used.

III is false because the effective stress is the pressure exerted by the soil particles on each other.

IV is false because increase in effective stress (rather than total stress) is responsible for consolidation settlement.

Answer is (A)

423

Converting each distance to miles and then calculating the product of Dia (in) × Length (miles), we have

Diameter (in)	Length (ft)	Length (mi)	DL (in-mi)
8	13,400	2.538	20.303
12	7,500	1.420	17.045
20	4,000	0.758	15.152
36	3,000	0.568	20.455
			72.955

Infiltration flow rate $= 100 \times 72.955 = 7,296$, gpd $= 5.07$, gpm $= 0.011$ cfs

Answer is (D)

424

Discharge $Q = 2,000$ gpm $= 4.464$ ft^3/sec

Hydraulic conductivity $K = 1,000$ ft/day $= 1.157 \times 10^{-2}$ ft/sec

At observation well 1, radial distance $r_1 = 30$ ft, height of piezometric surface (above aquifer bottom): $y_1 = 50 - 4.5 = 45.5$ ft

At observation well 2, radial distance $r_2 = 180$ ft, height of piezometric surface (above aquifer bottom): $y_2 = ?$

$$Q = \frac{\pi K\left(y_1^2 - y_2^2\right)}{\ln\left(\dfrac{r_1}{r_2}\right)} \Rightarrow \left(y_1^2 - y_2^2\right) = \frac{Q\ln\left(\dfrac{r_1}{r_2}\right)}{\pi K} = \frac{4.464 \times \ln\left(\dfrac{30}{180}\right)}{\pi \times 1.157 \times 10^{-2}} = -220$$

$y_2 = 47.9$ ft. Drawdown $s_2 = 50 - 47.9 = 2.1$ ft (A)

425

Time Period (hr)	Average Inflow Rate (ft³/sec)	Inflow × 10⁶ (gal)	I–O
00:00–02:00	18.3	0.986	0.407
02:00–04:00	12.4	0.668	0.089
04:00–06:00	8.7	0.469	–
06:00–08:00	7.3	0.393	–
08:00–10:00	6.4	0.345	–
10:00–12:00	8.9	0.479	–
12:00–14:00	14.5	0.781	0.202
14:00–16:00	18.9	1.018	0.439
16:00–18:00	6.5	0.350	–
18:00–20:00	5.6	0.302	–
20:00–22:00	8.3	0.447	–
22:00–24:00	13.2	0.711	0.132
		6.949	

Average flow per period $= 6.949/12 = 0.579$ Mgal

Streak with the largest cumulative surplus volume $(12:00 - 16:00) = 0.202 + 0.439 = 0.641$ Mgal $= 641,000$ gallons. (B)

426

For masonry construction $F = 0.8$

Fire flow demand: $Q = 18F\sqrt{A} = 18 \times 0.8 \times \sqrt{160,000} = 5,760$ gpm

Time for which the tank can provide this flow rate: $t = \dfrac{V}{Q} = \dfrac{700,000}{5,760} = 121.5$ min

Therefore, the tank meets fire flow demand for approximately 2 hr. (C)

427

Assuming that the primary sludge, which contains 6% solids, essentially has the same specific gravity as water, we can convert the 6% solids to a concentration:

Solids 6% (by weight) $= 60$ g/1,000 g $= 60,000$ mg/L (assuming sludge is "watery," 1 L weighs 1 kg $= 1,000$ g)

Primary effluent contains TSS = 0.35 × 250 = 87.5 mg/L

Influent contains 250 mg/L TSS. After 65% removal, primary effluent contains 0.35 × 250 = 87.5 mg/L TSS.

Performing a summation of mass flow rates at the node representing the primary clarifier, we get (where Q_{ps} is the volumetric flow rate of primary sludge in MGD):

$4 \times 250 = (4 - Q_{ps}) \times 87.5 + Q_{ps} \times 60{,}000 \Rightarrow Q_{ps} = 0.01085$ MGD = 10,850 gpd   (C)

428

$K = 3 \times 10^{-4}$ cm/sec = 1×10^{-5} ft/sec

For each side, $N_f = 3$, $N_e = 7$, $H = 16$ ft

Flow rate (per unit length) $q = 1 \times 10^{-5} \times 3/7 \times 16 = 6 \times 10^{-5}$ ft³/s = 4.11×10^{-3} ft³/min

Therefore, total seepage into the trench = $2 \times 4.11 \times 10^{-3}$ ft³/min/ft = 8.2×10^{-3} ft³/min/ft   (C)

429

$$DWEL = \frac{R_f D \times \text{Body weight}}{\text{Consumption}} = \frac{0.06 \times 70}{2} = 2.1 \frac{\text{mg}}{L}$$ (D)

430

Volume of trickling filter: $V = \dfrac{\pi}{4} \times 80^2 \times 6 = 30{,}159$ ft³

BOD load = 240 × 3 × 8.3454 = 6,008.7 lb/day

BOD loading rate: $L_{\text{BOD}} = \dfrac{6{,}008.7}{30.159} = 199.2 \dfrac{\text{lb}}{1{,}000 \text{ ft}^3\text{-day}}$

Recirculation factor: $F = \dfrac{1 + R}{(1 + 0.1R)^2} = \dfrac{1 + 3}{(1 + 0.3)^2} = 2.37$

Removal efficiency: $\eta = \dfrac{1}{1 + 0.0561\sqrt{\dfrac{199.2}{2.37}}} = 0.66$

Therefore, the BOD_5 in the effluent = 34% of 240 = 81.6 mg/L   (B)

431

Let us assume that the factory can discharge a maximum lead concentration $= x$ (g/L)

Wastewater flow rate $= 2$ MGD $= 2 \times 1.5472 = 3.094$ ft³/sec

The lead concentration in the stream immediately after mixing:

$$\overline{Pb} = \frac{3.094x + 30 \times 4 \times 10^{-6}}{3.094 + 30}$$

Since the EPA limit on [Pb] is 15 µg/L, $\overline{Pb} = \dfrac{3.094x + 30 \times 4 \times 10^{-6}}{3.094 + 30} \le 15 \times 10^{-6} \Rightarrow$

$x \le 121 \times 10^{-6}$

Therefore, the plant must reduce its lead emission of 0.5 mg/L (500 µg/L) to 121 µg/L.

This is a removal rate of 76% **(A)**

432

Time for stream to travel 20 mi (105,600 ft) at 4 ft/sec $= 26,400$ sec (7.33 hr)

Assuming that the stream returns to its equilibrium temperature of 12°C within a short distance, there is no need to calculate the weighted average temperature at the instant of mixing.

At temperature of 12°C, saturation D.O. $= 10.77$ mg/L (table)

At a distance of 20 mi (7.33 hr away), the dissolved oxygen concentration $= 4.4$ mg/L

Therefore, 5 mi downstream, the oxygen deficit $= 10.77 - 4.4 = 6.37$ mg/L **(B)**

433

At 5 days, the BOD is expected to be on the exponential BOD growth curve. Therefore, the ultimate BOD is calculated from:

$$BOD_t = BOD_{ult}(1 - 10^{-kt}) \Rightarrow BOD_{ult} = \frac{4.5}{(1 - 10^{-0.1 \times 5})} = 6.58 \text{ mg/L}$$

According to the same model, the 20-day (carbonaceous BOD would have been)
$BOD_{20} = 6.58 \times (1 - 10^{-0.1 \times 20}) = 6.52$ mg/L

At 20 days, the excess (recorded) BOD comes from nitrogenous bacteria.

Nitrogenous BOD = 8.3 − 6.52 = 1.78 mg/L   (D)

434

Among those listed in the water analysis, the contaminants affecting drinking water standards are nitrate, turbidity, odor, total coliform, and total dissolved solids (TDS).

Nitrate concentration = 21 mg/L, converted to NO_3-N = 21 × 14/62 = 4.74 mg/L is within limits (10 mg/L) of primary DWS.

Turbidity exceeds limits of primary DWS (1.3 NTU > 1 NTU).

Odor is within limits (2.7 < 3.0) of secondary DWS.

Total coliform exceeds limits of primary DWS (1.7 MPN > 1.0 MPN).

TDS is within limits (437 < 500) of secondary DWS.

Therefore, turbidity and total coliform exceed drinking water standards.   (D)

435

Dissolved solids are those that pass through the filter in solution. The volatile dissolved solids will be those that are burnt off upon ignition.

In a 200 mL (0.2 L) solution, VDS = 47.225 − 46.201 = 1.024 g = 1,024 mg

VDS concentration = 1,024/0.2 = 5,120 mg/L   (D)

436

The definition of pH is logarithmic, as shown below:

$pH = -\log_{10}[H^+]$

where $[H^+]$ is the concentration of the H^+ ion in moles per liter.

Thus, before averaging the two flows, the pH of each must be converted to these concentrations, which can then be averaged.

(*Continued on next page*)

For the stream, $[H^+] = 10^{-7.8} = 1.585 \times 10^{-8}$ moles/L

For the wastewater, $[H^+] = 10^{-3.4} = 3.98 \times 10^{-4}$ moles/L; and flow rate $= 2,500/694.4 = 3.6$ mgd

The average concentration is: $[H^+] = \dfrac{5 \times 1.585 \times 10^{-8} + 3.6 \times 3.98 \times 10^{-4}}{5 + 3.6} = 1.667 \times 10^{-4}$

Average pH, assuming perfect mixing $= -\log_{10}(1.667 \times 10^{-4}) = 3.78$ **(C)**

437

Average velocity: $V = \dfrac{Q}{A} = \dfrac{2,514}{1,323} = 1.9 \ \dfrac{ft}{sec}$

Reaeration coefficient: $k_r = \dfrac{3.3V}{H^{1.33}} = \dfrac{3.3 \times 1.9}{3.8^{1.33}} = 1.06$ **(C)**

438

Chlorine residual $=$ dose $-$ demand $= 5.0 - 3.5 = 1.5$ mg/L

At this concentration, for 2-log inactivation, $CT = 28$ mg/L-min

Therefore, contact time $T = 28/1.5 = 18.67$ min $= 0.013$ day

Volume of chlorination chamber: $V = QT = 3.5 \times 10^6 \times 0.013 = 45,370$ gal $= 6,065$ ft^3 **(A)**

439

The "as species" concentrations of the divalent metal ions (Ca^{++}, Mg^{++}, and Fe^{++}) should be converted to "as $CaCO_3$" equivalents utilizing the factors calculated from equivalent weight.

$[Ca^{++}] = 60.0 \times 2.5 = 150$ mg/L as $CaCO_3$

$[Mg^{++}] = 21.2 \times 4.1 = 86.9$ mg/L as $CaCO_3$

$[Fe^{++}] = 2.2 \times 1.77 = 3.9$ mg/L as $CaCO_3$

Hardness $= 240.8$ mg/L as $CaCO_3$ **(A)**

440

For a perpetual annuity, capitalized cost $= A/i$

All costs are calculated in thousands of dollars.

Present worth of Plan A (single capital investment) is calculated as the sum of initial expense + present worth of perpetual annuity.

$$P_A = 420 + 40 \times \frac{1}{0.07} = 991.43$$

Present worth of Plan B (two-stage capital investment) is calculated as the sum of initial expense + lump-sum expense at $t = 10$ years + perpetual annuity of 16 + 10-year annuity of 4

$$P_A = 200 + 320 \left(\frac{P}{F}, 10 \text{ years}, 7\% \right) + 16 \times \frac{1}{0.07} + 4 \left(\frac{P}{A}, 10 \text{ years}, 7\% \right)$$

$$= 200 + 320 \times 0.5083 + 16 \div 0.07 + 4 \times 7.0236 = 619.322$$

Cost ratio $= 619.322 \div 991.43 = 0.625$ (A)

Answer Key for Water Resources & Environmental Depth Exam

401	B	411	B	421	D	431	A
402	C	412	B	422	A	432	B
403	C	413	C	423	D	433	D
404	B	414	B	424	A	434	D
405	D	415	B	425	B	435	D
406	B	416	D	426	C	436	C
407	A	417	A	427	C	437	C
408	C	418	C	428	C	438	A
409	D	419	A	429	D	439	A
410	D	420	A	430	B	440	A

13

Transportation Depth Exam Solutions

These detailed solutions are for questions 501 to 540, representative of a 4-hr Transportation Depth exam according to the syllabus and guidelines for the Principles and Practice (P&P) of Civil Engineering Examination administered by the National Council of Examiners for Engineering and Surveying (NCEES), current for the October 2020 examination.

501

According to the Gravity model, the number of trips attracted from zone i to zone j is given in terms of trip production of zone $i(P_i)$ and trip attractions of zone $j(A_j)$ by:

$$T_{ij} = P_i\left[\frac{A_j F_{ij}}{\sum A_j F_{ij}}\right] = 440 \times \left[\frac{350 \times 35}{120 \times 75 + 350 \times 35 + 670 \times 40}\right] = 112 \tag{A}$$

502

According to the HCM 6th edition, the lane capacity (pc/h) is given by:

$$c_{c,pce} = 1{,}130e^{-1.0\times10^{-3}\,v_{c,pce}} = 1{,}130 \times e^{-0.001\times230} = 898 \tag{C}$$

503

Directional ADT $= 0.6 \times 65{,}000 = 39{,}000$ veh/day

Design hourly volume: DHV $= K \cdot ADT = 0.12 \times 39{,}000 = 4{,}680$ veh/day

For 12-ft lanes, $f_{LW} = 0.0$

For 8-ft shoulders, $f_{LC} = 0.0$

A full cloverleaf interchange has four ramps, TRD $= 4$ ramps per 1.25 mi $= 3.2$ ramps per mi

$$FFS = 75.4 - f_{LC} - f_{LW} - 3.22 \, TRD^{0.84} = 75.4 - 0 - 0 - 3.22 \times 3.2^{0.84} = 66.85 \text{ mph}$$

$$f_{HV} = \frac{1}{1 + P_T \left(E_T - 1 \right)} = \frac{1}{1 + 0.16 \times \left(2.0 - 1 \right)} = 0.862$$

$$v_p = \frac{V}{N f_P \, f_{HV} \, PHF} = \frac{4680}{3 \times 1.0 \times 0.862 \times 0.9} = 2011$$

For $v_p = 2011$ pcphpl and FFS $= 66.85$ mph, using HCM equations, speed $S = 60.7$ mph.

Density $D = 33.1$ pcpmpl

LOS D   (D)

504

Lane width $= 11$-ft lanes. From HCM, $f_{LW} = 1.9$ mph

Total lateral clearance $= LC_L + LC_R = 4 + 6 = 10$ ft. From HCM, $f_{LC} = 0.4$ mph

For divided highway with median, HCM gives $f_M = 0.0$ mph

Access point density $= 5{,}280/600 = 8.8$ access points per mile. From HCM, $f_A = 2.2$ mph

Default BFFS $=$ Speed limit $+ 5$ mph $= 55$ mph

FFS $=$ BFFS $- f_{LW} - f_{LC} - f_M - f_A = 55 - 1.9 - 0.4 - 0 - 2.2 = 50.5$ mph

The traffic stream includes 8% trucks, 3% buses, and 2% RVs. Heavy vehicle factor is:

$$f_{HV} = \frac{1}{1 + P_T \left(E_T - 1 \right)} = \frac{1}{1 + 0.13 \times \left(2.0 - 1 \right)} = 0.885$$

Peak flow rate is calculated as:

$$v_p = \frac{V}{Nf_P f_{HV} PHF} = \frac{3,440}{3 \times 1.0 \times 0.885 \times 0.88} = 1,472$$

LOS D (approximate range: 1,300 to 1,700)   (B)

505

Segment capacity = 3,200 pc/h (for both directions) or 1,700 pc/h for each direction.

Flow rate (both directions) = 2,800, does not exceed 3,200

Flow rate (design direction) = 0.6 × 2,800 = 1,680, does not exceed 1,700.

From exhibit 15-3 of the Highway Capacity Manual 6th edition, for average travel speed (ATS) = 42 mph, level of service is D; for percent time spent following (PTSF) = 63, level of service is C.

Governing LOS = D   (C)

506

Actual space-hour utilization is calculated as the sum $\Sigma N_i h_i$, where N_i is the number of cars in a particular group and h_i is the number of hours these cars are parking for.

$$\sum N_i h_i = 360 \times (0.1 \times 1 + 0.15 \times 2 + 0.2 \times 3 + 0.3 \times 4 + 0.25 \times 10) = 1,692 \text{ space} \cdot \text{hr}$$

This represents 85% occupancy. Therefore, space hours available = 1,692 ÷ 0.85 = 1,990

Assuming N spaces in the lot, with the lot operating 10 hr a day, with 80% efficiency, number of space-hours available = 0.8 × 10 × N = 8N, which must be equal to 1,990. Thus, N = 248.8.   (A)

507

Jam density $D_j = 64$ veh/mi

Free flow speed $S_f =$ twice the optimum speed $= 100$ mph

Capacity, $c = \dfrac{1}{4} S_f D_j = \dfrac{1}{4} \times 100 \times 64 = 1600$ veh/hr **(B)**

Note: Greenshield's (linear) model has been used here.

$$S = S_f \left(1 - \frac{D}{D_j} \right)$$

508

Number of entering vehicles (24 hr) $= 5,720$ (ADT)

Accident rate: $R = \dfrac{\text{No. of accidents} \times 10^8}{\text{ADT} \times 365 \times N} = \dfrac{15 \times 10^8}{5,720 \times 365 \times 1} = 718$

Accident rate $= 718$ accidents per hundred million entering vehicles **(C)**

509

ADT in 2010 $= 15,600 \times 1.03^2 = 16,550$

Crashes expected (without crash modifications) in 2010 $= 12 \times 1.03^2 = 12.73$

Overall crash reduction factor:

$$CR = CR_1 + (1 - CR_1)CR_2 + (1 - CR_1)(1 - CR_2)CR_3$$
$$= 0.1 + 0.9 \times 0.25 + 0.9 \times 0.75 \times 0.2 = 0.46$$

Number of crashes prevented $= 12 \times 0.46 \times 16,550/15,600 = 5.86$

Number of crashes expected in 2010 $= 12.73 - 5.86 = 6.87$ **(B)**

510

For acceleration phase, time $= 70 \div 8 = 8.75$ sec

Acceleration rate $= 8$ mph/sec $= 11.76$ ft/sec^2

Distance $= \frac{1}{2}\, at^2 = 0.5 \times 11.76 \times 8.75^2 = 450.2$ ft

For deceleration phase, time = 70 ÷ 10 = 7.0 sec

Deceleration rate = 10 mph/sec = 14.7 ft/sec²

Distance = ½ at² = 0.5 × 14.7 × 7² = 360.2 ft

Since total distance traveled = 0.5 mi = 2,640 ft, this leaves distance for constant velocity phase = 2,640 − 450.2 − 360.2 = 1,829.6 ft

Time for constant velocity phase = 1829.6 ÷ (1.47 × 70) = 17.8 sec

Total travel time = 8.75 + 17.8 + 7.0 = 33.53 sec

Average running speed = 2,640 ÷ 33.53 = 78.73 fps = 53.6 mph　　　　　　　　　(C)

511

The composite value of the pedestrian space is calculated (using a harmonic average) as:

$$A_{p,F} = \frac{\sum L_i}{\sum \left(\dfrac{L_i}{A_{p,i}} \right)} = \frac{410}{\dfrac{25}{10} + \dfrac{150}{50} + \dfrac{210}{30} + \dfrac{25}{12}} = 28.1 \qquad (D)$$

512

Degree of curve = 4. Therefore, curve radius $R = 5{,}729.578/D = 1{,}432.4$ ft

Tangent length: $T = R\tan\dfrac{I}{2} = 768.25$ ft

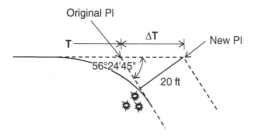

The tangent length will get longer by $\Delta T = \dfrac{20}{\sin 56°24'45''} = 24.01$ ft

(Continued on next page)

New tangent length, $T + \Delta T = 768.25 + 24.01 = 792.26$ ft

New radius, $R' = \dfrac{T'}{\tan\dfrac{I}{2}} = \dfrac{792.26}{\tan 28.21} = 1{,}477.17$ ft

New degree of curve, $D' = \dfrac{5{,}729.578}{1477.17} = 3.88$ (C)

513

The back tangent has bearing N 46°25′32″E. Since this is in the NE quadrant, this is also its azimuth. The forward tangent has bearing S 17°56′21″E. Since this is in the SE quadrant, its azimuth $= 180° - 17°56′21″ = 162°3′39″$

The deflection angle is the difference of the azimuths $= 162°3′39″ - 46°25′32″ = 115°38′07″$ (115.635°)

Length of back tangent $T_a =$ distance from PC to PI $= 15{,}812.98 - 13{,}834.12 = 1{,}978.86$ ft

$$T_a = \frac{R_2 - R_1\cos I + (R_1 - R_2)\cos\Delta_2}{\sin I}$$

$$1{,}978.86 = \frac{900 - 1{,}800 \times \cos 115.635 + (1{,}800 - 900)\cos\Delta_2}{\sin 115.635} \Rightarrow \Delta_2 = 83.279, \Delta_1 = 32.356$$

$$L_1 = 100\frac{\Delta_1}{D_1} = 100\frac{\Delta_1 R_1}{5{,}729.578} = 1{,}016.49 \text{ ft}$$

$PCC = PC + L_1 = 138 + 34.12 + 10 + 16.49 = 148 + 50.61$ (C)

514

Azimuth of back tangent $= 180° + 42°30′ = 222°30′$

Azimuth of forward tangent $= 360° - 70° = 290°00′$

Deflection angle between tangents, $I = 290°00′ - 222°30′ = 67°30′$

Radius: $R = \dfrac{5{,}729.578}{D} = \dfrac{5{,}729.578}{3.75} = 1{,}527.89$ ft

Length of curve: $L = \dfrac{100I}{D} = \dfrac{100 \times 67.5}{3.75} = 1800$ ft

Tangent length: $T = R\tan\dfrac{I}{2} = 1{,}527.89 \times \tan\dfrac{67.5}{2} = 1{,}020.90$ ft

PI station: $50 + 22.30$

PC station $=$ PI station $- T = (50 + 22.30) - (10 + 20.90) = 40 + 01.40$

Arc length from PC to station $57 + 00.00 = 1{,}698.6$ ft

Deflection angle: $\alpha = \text{arc} \times \dfrac{I}{2L} = 1{,}698.6 \times \dfrac{67.5}{2 \times 1{,}800} = 31.849° = 31°50'56''$ \quad (A)

515

Travel Segment 1: On Highway – Initial speed $= 65$ mph (95.3 fps). Final speed unknown

Braking deceleration on highway $= fg = 0.29 \times 32.2 = 9.34$ ft/s^2

Braking distance: $70 = \dfrac{V_i^2 - V_f^2}{2g(f - G)} = \dfrac{95.3^2 - V_f^2}{2 \times 32.2(0.29 - 0.06)} \Rightarrow V_f = 89.7$ fps

Travel Segment 2: On Ramp – Initial speed $= 89.7$ fps. Final speed $= 0$. Distance unknown

Braking deceleration on ramp $= fg = 0.4 \times 32.2 = 12.88$ ft/s^2

Braking distance: $d = \dfrac{V_i^2 - V_f^2}{2g(f + G)} = \dfrac{89.7^2 - 0}{2 \times 32.2(0.4 + 0.1)} = 249.9$ ft \quad (C)

516

Distance from PVC_2 to $\text{PVI}_2 = 36.6965 - 28.2238 = 8.4727$ sta. This represents half the length of curve. Therefore, length of curve 2 $= 16.9454$ sta

Rate of gradient change for curve 2, $R = \dfrac{9}{16.9454} = 0.53112\%/\text{sta}$

Elevation of $\text{PVC}_2 =$ elev $\text{PVI}_2 - G_1\dfrac{L}{2} = 371.12 - (-5) \times 8.4727 = 413.48$ ft

Distance to bridge structure from $\text{PVC}_2 = 40.55 - 28.2238 = 12.3262$ sta

(Continued on next page)

Curve elevation at $40 + 55$ is:

$$y = 413.48 + (-5) \times 12.3262 + \frac{1}{2} \times 0.53112 \times 12.3262^2 = 392.2 \text{ ft}$$

Vertical clearance $= 405.54 - 392.20 = 13.34$ ft

For a sag curve with limited clearance, AASHTO Green Book gives:

$$\text{For } S > L, \qquad L = 2S - \frac{800}{A}\left(C - \frac{h_1 + h_2}{2} \right)$$

Given $C = 13.34$, $h_1 = 8.0$, $h_2 = 2.0$, $L = 1694.54$ ft, $A = 9$, this equation yields $S = 1219$ ft. This solution does not satisfy the criterion $(S > L)$

$$\text{For } S \leq L, \qquad L = \frac{AS^2}{800\left(C - \dfrac{h_1 + h_2}{2} \right)}$$

Given $C = 13.34$, $h_1 = 8.0$, $h_2 = 2.0$, $L = 1694.54$ ft, $A = 9$, this equation yields $S = 1121$ ft. This solution satisfies the criterion $(S \leq L)$.

Corresponding design speed $= 90$ mph (impractical, of course, but solely based on sight distance under overhead obstruction) **(D)**

517

Distance from PVC to PVI $= 6{,}009.0 - 5{,}312.50 = 696.50$ ft

Length of curve $= 2 \times 696.50 = 1{,}393$ ft $= 13.93$ stations

Rate of grade change: $R = \dfrac{G_2 - G_1}{L} = \dfrac{3 - (-5)}{13.93} = 0.5743\%/\text{sta}$

Elevation of PVC: $y_{PVC} = y_{PVI} - G_1\dfrac{L}{2} = 365.57 - (-5) \times 6.965 = 400.40$ ft

Bridge (at sta $55 + 05.20$) is located at $x = 55.052 - 53.125 = 1.927$ stations

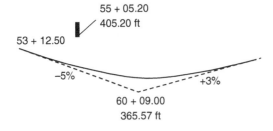

55 + 05.20
405.20 ft
53 + 12.50
−5%
+3%
60 + 09.00
365.57 ft

Curve elevation at location of bridge:

$$y = y_{\text{PVC}} + G_1 x + \frac{1}{2} R x^2 = 400.40 + (-5) \times 1.927 + \frac{1}{2} \times 0.5743 \times 1.927^2 = 391.83 \text{ ft}$$

Vertical clearance = $405.20 - 391.83 = 13.37$ ft   (B)

518

$$A = |G_2 - G_1| = 3.6\%$$

Design speed = 65 mph. Corresponding stopping sight distance (AASHTO Green Book) = reaction time distance (239 ft) + braking distance (405 ft) = 644 ft

Assuming $S < L$, $L = \dfrac{AS^2}{2,158} = \dfrac{3.6 \times 644^2}{2,158} = 692$ ft

This solution fits the criterion $S < L$

Therefore, the solution is OK. Minimum length of curve = 692 ft   (D)

519

Rate of gradient change: $R = \dfrac{G_2 - G_1}{L} = \dfrac{5 - (-3)}{9.5} = 0.842\%/\text{sta}$

Lowest point located at sta $x = -\dfrac{G_1}{R} = -\dfrac{-3}{0.842} = 3.563$ sta

Elevation at low point:

$$y = y_{\text{PVC}} + G_1 x + \frac{1}{2} R x^2 = 247.65 - 3 \times 3.563 + \frac{1}{2} \times 0.842 \times 3.563^2 = 242.31 \text{ ft}$$

Vertical clearance = $242.31 - 238.23 = 4.08$ ft   (D)

520

AASHTO Green Book, 7th edition, 2018, Chapter 9.

From exhibit 9-6, for left turn from Stop on minor street, where the major street has four lanes, base critical gap for a single unit truck = 9.5 sec.

Adjustment factor for additional lane crossed = 0.5 sec for four-lane major street

Adjustment factor for 4% grade on minor, $t_{c,G} = 4 \times 0.2 = 0.8$ sec for left turn from minor

Adjusted time gap = $9.5 + 0.5 + 0.8 = 10.8$ sec   (A)

521

Converting all hourly volumes to flow rates, we have:

$$v_{FF} = \frac{V_{FF}}{PHF} = \frac{1780}{0.88} = 2{,}023$$

$$v_{FR} = \frac{V_{FR}}{PHF} = \frac{366}{0.88} = 416$$

$$v_{RF} = \frac{V_{RF}}{PHF} = \frac{454}{0.95} = 478$$

$$v_{RR} = \frac{V_{RR}}{PHF} = \frac{580}{0.95} = 611$$

Weaving demand flow rate: $v_W = v_{FR} + v_{RF} = 894$

Non-weaving demand flow rate: $v_{NW} = v_{FF} + v_{RR} = 2{,}634$

$$\text{Volume ratio: VR} = \frac{v_W}{v_W + v_{NW}} = \frac{894}{894 + 2634} = 0.253 \qquad \text{(A)}$$

522

Upstream peak flow rate, $V_F = 4{,}100 + 500 = 4{,}600$

Adjusted flow entering lanes 1 and 2 immediately upstream of diverge influence area is calculated as:

$$V_{12} = V_R + (V_R - V_R)P_{FD}$$

For six-lane freeways, there are three possible equations for P_{FD}. No upstream ramp and an adjacent downstream off-ramp, equation 7 or 5 applies. Calculate L_{EQ}

$$L_{EQ} = \frac{V_D}{1.15 - 0.000032V_F - 0.000369V_R}$$
$$= \frac{450}{1.15 - 0.000032 \times 4600 - 0.000369 \times 500} = 550 \text{ ft}$$

Since distance to adjacent downstream ramp is greater than L_{EQ}, use equation 5

$$P_{\text{FD}} = 0.76 - 0.000025 V_F - 0.000046 V_R = 0.76 - 0.000025 \times 4600 - 0.000046 \times 500$$
$$= 0.622$$

$$V_{12} = 500 + (4600 - 500) \times 0.622 = 3050 \text{ pcph} \tag{C}$$

523

According to the MUTCD, the minimum warning distance is 175 ft (Table 2C-4)   (C)

Table 2C-4. Guidelines for Advance Placement of Warning Signs

Posted or 85th-Percentile Speed	Condition A: Speed reduction and lane changing in heavy traffic[2]	Advance Placement Distance[1]							
		Condition B: Deceleration to the listed advisory speed (mph) for the condition							
		0[3]	10[4]	20[4]	30[4]	40[4]	50[4]	60[4]	70[4]
20 mph	225 ft	100 ft[6]	N/A[5]	—	—	—	—	—	—
25 mph	325 ft	100 ft[6]	N/A[5]	N/A[5]	—	—	—	—	—
30 mph	460 ft	100 ft[6]	N/A[5]	N/A[5]	—	—	—	—	—
35 mph	565 ft	100 ft[6]	N/A[5]	N/A[5]	N/A[5]	—	—	—	—
40 mph	670 ft	125 ft	100 ft[6]	100 ft[6]	N/A[5]	—	—	—	—
45 mph	775 ft	175 ft	125 ft	100 ft[6]	100 ft[6]	N/A[5]	—	—	—
50 mph	885 ft	250 ft	200 ft	175 ft	125 ft	100 ft[6]	—	—	—
55 mph	990 ft	325 ft	275 ft	225 ft	200 ft	125 ft	N/A[5]	—	—
60 mph	1,100 ft	400 ft	350 ft	325 ft	275 ft	200 ft	100 ft[6]	—	—
65 mph	1,200 ft	475 ft	450 ft	400 ft	350 ft	275 ft	200 ft	100 ft[6]	—
70 mph	1,250 ft	550 ft	525 ft	500 ft	450 ft	375 ft	275 ft	150 ft	—
75 mph	1,350 ft	650 ft	625 ft	600 ft	550 ft	475 ft	375 ft	250 ft	100 ft[6]

524

For ADT = 16,000, design speed = 65 mph and foreslope = 1:4, width of clear zone = 46 mph. This includes the shoulder width. Therefore, clear distance from edge of shoulder = 46 − 8 = 38 ft   (D)

525

Traffic barriers are sometimes used to separate bicycles and other slower moving modes from vehicular traffic streams. They are also used to separate opposing directions of vehicular traffic and to provide a safe haven to workers and equipment in a work zone adjacent to traffic lanes.

Even though the use of barriers may result in more streamlined traffic movement and therefore, capacity increases, their design is NOT ostensibly based on that objective   (A)

526

WARRANT 1: Eight-Hour Vehicular Volume

Warrant 1: Condition A thresholds (100% level) for 1-lane major − 1-lane minor are 500 vph total on both approaches major *and* 150 vph on higher volume minor. These thresholds are exceeded during only 7 hr. So condition A is *not met* at 100% level

Warrant 1: Condition B thresholds (100% level) for 1-lane major − 1 lane minor are 750 vph total on both approaches major *and* 75 vph on higher volume minor. The two criteria for Condition B of Warrant 1 are met during 11 hr.

Warrant 1 requires the thresholds met during any 8 hr. So, this condition of warrant 1 is met.

Overall, Warrant 1 is met if *either* condition A or condition B is met at the 100% level.

WARRANT 2: Four-Hour Vehicular Volume

The data in the table is plotted on Fig. 4C-1. All but one data point plots above the line for 1-lane major and 1-lane minor. The warrant requires that this threshold be exceeded for at least 4 hr on a single day. Therefore, this warrant is met.

Answer is   (B)

527

According to guidelines in the Americans with Disabilities Act Accessibility Guidelines, when the total parking in the lot is from 501 to 1,000, the minimum number of accessible spaces shall be 2% of the total. For a total number of spaces = 630, this results in 12.6 spaces. Provide 13 spaces.   (D)

528

Phase	Lane Group Volume	Saturation Volume for Lane Group
1	100	900
2	600	1,900
3	105	900
4	630	2,300

Total lost time $= 12$ sec

Webster's formula gives the optimum cycle length as

$$C = \frac{1.5t_L + 5}{1 - \sum Y_i} = \frac{1.5 \times 12 + 5}{1 - \left(\dfrac{100}{900} + \dfrac{600}{1900} + \dfrac{105}{900} + \dfrac{630}{2300}\right)} = \frac{23}{1 - 0.8175} = 126 \text{ sec} \qquad \text{(B)}$$

529

$$f_{HV} = \frac{1}{1 + P_{HV}(E_{HV} - 1)} = \frac{1}{1 + 0.05 \times (2.0 - 1)} = 0.952$$

With 25 parking maneuvers/hr, $f_p = \dfrac{N - 0 \cdot 1 - \dfrac{18N_m}{3600}}{N} = \dfrac{2 - 0 \cdot 1 - \dfrac{18 \times 25}{3600}}{2} = 0.888$

With 10 buses/hr, $f_{bb} = \dfrac{N - \dfrac{14.4N_B}{3600}}{N} = \dfrac{2 - 14.4 \times \dfrac{10}{3,600}}{2} = 0.98$

Saturation flow rate on WBTH/R approach (assuming non-CBD area)

$$s = s_0 N f_w f_{HV} f_g f_p f_{bb} f_a f_{RT} f_{LT} = 1900 \times 2 \times 1.0 \times 0.952 \times 0.995 \times 0.888 \times$$
$$0.98 \times 1.0 \times 0.9625 \times 0.95 = 2864 \text{ pcph} \qquad \text{(B)}$$

530

The clearance interval (for the wider pavement width $= 60$ ft) is calculated as

$$\tau_{min} = t_R + \frac{V}{2a} + \frac{W + L}{V} = 1.0 + \frac{1.47 \times 40}{2 \times 10} + \frac{60 + 20}{1 \cdot 47 \times 40} = 5.3 \text{ sec} \qquad \text{(C)}$$

531

Length of walkway $= 60$ ft

Cycle time $= 80$ sec; number of cycles per hour $= 3,600/80 = 45$

Number of pedestrians per cycle $N_{ped} = 1200 \div 45 = 26.67$

Green time: $G_p = 3.2 + \dfrac{L}{S_p} + 0.27N_{ped} = 3.2 + \dfrac{60}{4.0} + 0.27 \times 26.67 = 25.4 \text{ sec} \qquad \text{(A)}$

532

Warrant 1:   Condition A – 600 major and 150 on minor

Condition B – 900 major and 75 on minor

Neither is met for ALL 8 hr at 100% level.

Both are not met for all 8 hr at 80% level.

Therefore, warrant 1 is not met.

Warrant 2: Is met for hours 4, 6, 7, and 8.

Warrant 3: Meets condition (2). Point plots above line for hr 7.

Warrants 2 and 3 are met.   (D)

533

Answer is A (MUTCD section 6G.02)   (A)

534

$F_{200} = 8$, LL $= 43$, PI $= 43 - 21 = 22$

According to the chart below, the criteria: $F_{200} < 35$, LL > 40, and
PI > 10 match group A-2-7   (D)

Table 203.5 AASHTO Soil Classification Criteria

Sieve Analysis	Granular Materials (35% or less passing no. 200 sieve)							Silt-Clay Materials (more than 35% passing no. 200 sieve)				A-8
	A-1			A-2								
	A-1-a	A-1-b	A-3	A-2-4	A-2-5	A-2-6	A-2-7	A-4	A-5	A-6	A-7	A-8
% passing												
No. 10	≤50											
No. 40	≤30	≤50	>50									
No. 200	≤15	≤25	≤10	≤35	≤35	≤35	≤35	>35	>35	>35	>35	
LL				≤40	>40	≤40	>40	≤40	>40	≤40	>40	
PI	≤6		NP	≤10	≤10	>10	>10	≤10	≤10	>10	>10	

535

By examination, it seems that the peak dry unit weight will come from sample 3 or 4.

Volume of Standard Proctor mold $= 1/30 \text{ ft}^3$

For sample 3, total unit weight, $\gamma = W/V = 3.95/(1/30) = 118.5 \text{ lb/ft}^3$

Dry unit weight, $\gamma_d = 118.5/1.16 = 102.2 \text{ lb/ft}^3$

For sample 4, total unit weight, $\gamma = W/V = 4.21/(1/30) = 126.3 \text{ lb/ft}^3$

Dry unit weight, $\gamma_d = 126.3/1.18 = 107.0 \text{ lb/ft}^3$ (maximum dry unit weight)

Target dry unit weight $= 0.9 \times 107 = 96.3 \text{ lb/ft}^3$

Weight of solids needed $= 96.3 \times 1.5 \times 10^6 \times 27 = 3.9 \times 10^9 \text{ lb}$

From table, for 12% moisture content, unit weight $= 3.24/(1/30) = 97.2 \text{ lb/ft}^3$; $\gamma_d = 97.2/1.12 = 86.8 \text{ lb/ft}^3$

Volume of borrow soil needed $=$
$3.9 \times 10^9 \text{ lb} \div 86.8 \text{ lb/ft}^3 = 4.49 \times 10^7 \text{ ft}^3 = 1.66 \times 10^6 \text{ yd}^3$ **(B)**

536

Number of trucks $= 23{,}000 \times 0.12 = 2{,}760$ per day

Axle load data is based on 1,078 trucks. Therefore, results must be scaled by factor $2{,}760/1{,}078 = 2.56$

Growth percentage $= 4\%$

Annual ADT numbers represent a geometric series (20 terms, $r = 1.04$). The cumulative ADT over 20 years is the sum of the series given by

$$S = a\left(\frac{r^n - 1}{r - 1}\right) = \text{ADT}_1\left(\frac{1.04^{20} - 1}{1.04 - 1}\right) = 29.778 \times \text{ADT}_1$$

The year 1 daily ESAL:

$$W_{18} = \sum N_i LEF_i = 2420 \times 0.0877 + 630 \times 0.36 + 301 \times 1 + 22 \times 1.51 + 6 \times 2.18 +$$
$$1 \times 3.53 + 24 \times 0.18 + 15 \times 0.308 + 12 \times 0.495 + 11 \times 0.857 = 814$$

Cumulative 20-year ESAL: $W_{18} = 814 \times 365 \times 29.778 \times 2.56 = 22.65$ million **(D)**

537

I is a true statement. The number of punchouts per mile, a predictor of pavement performance is very sensitive to pavement thickness. In some cases, a reduction of only ¼ in in PCC thickness results in a near-doubling of the CRCP punchouts.

II is not true. Punchouts are caused by cyclic *tensile* stresses between the wheels in the upper layers of the pavement.

III is a true statement. The base type selected for support in a CRCP is a critical factor impacting projected performance both in the development of cracks and tight crack widths as well as in resisting foundation layer erosion from repeated loading.

IV is a true statement. The MEPDG uses a fatigue model to model the effect of repeated load cycles on the development of cracks

V is not true. The iterative nature of the MEPDG procedure allows the design engineer limits the allowable number of punchouts at the end of the design life to an acceptable level (typically between 10 and 20 per mile) at a given level of reliability.

Answer is (B)

538

After the base-flow is subtracted, we get:

Time (hr)	0	1	2	3	4	5	6
Discharge Q (ft³/sec)	0	61	104	89	52	9	2

Volume of runoff: $V = \Sigma Q \times \Delta t = 317 \times 3,600 = 1,141,200$ ft³

Average depth of runoff (excess precipitation) = $1,141,200/(115 \times 43,560) = 0.228$ ft = 2.73 in.

The peak **runoff** discharge (ft³/sec) that would be recorded following a 2-hr storm producing 1.7 in of runoff is most nearly: $104 \times 1.7/2.73 = 65$ ft³/sec

Adding in the base flow (unaffected by storm), the peak discharge that would be recorded in the stream = 88 ft³/sec (A)

539

For pipe diameter = 36 in and Q = 100 cfs, FHWA nomograph yields HW/D = 3.5

Headwater = $3.5 \times 3 = 10.5$ ft (this is measured from the upstream invert elevation)

100-year water surface elevation = 325 + 10.5 = 335.5 ft

Vertical clearance between roadway and 100-year WSE = 337.8 − 335.5 = 2.3 ft   **(B)**

540

For a perpetual annuity, capitalized cost = A/i

All costs are calculated in millions of dollars.

Present worth of Plan A (single capital investment) is calculated as the sum of initial expense + present worth of perpetual annuity.

$$P_A = 4.2 + 0.04 \times \frac{1}{0.07} = 4.771$$

Present worth of Plan B (two-stage capital investment) is calculated as the sum of initial expense + lump-sum expense at t = 10 years + perpetual annuity of 16 + 10-year annuity of 4

$$P_B = 2.0 + 3.2 \left(\frac{P}{F}, 10 \text{ yrs}, 7\% \right) + 0.036 \times \frac{1}{0.07} - 0.008 \left(\frac{P}{A}, 10 \text{ yrs}, 7\% \right)$$
$$= 2.0 + 3.2 \times 0.5083 + 0.036 \div 0.07 - 0.008 \times 7.0236 = 4.085$$

Cost ratio = 4.085 ÷ 4.771 = 0.856   **(A)**

Answer Key for Transportation Depth Exam

501	A		511	D		521	A		531	A
502	C		512	C		522	C		532	D
503	D		513	C		523	C		533	A
504	B		514	A		524	D		534	D
505	C		515	C		525	A		535	B
506	A		516	D		526	B		536	D
507	B		517	B		527	D		537	B
508	C		518	D		528	B		538	A
509	B		519	D		529	B		539	B
510	C		520	A		530	C		540	A

14

Construction Depth Exam Solutions

These detailed solutions are for questions 601 to 640, representative of a 4-hr Construction Depth exam according to the syllabus and guidelines for the Principles and Practice (P&P) of Civil Engineering Examination administered by the National Council of Examiners for Engineering and Surveying (NCEES), current for the October 2020 examination.

601

Using the average end area method, and adjusting the fill volumes for shrinkage, we get the following:

Station	Area (ft²)		Volume (yd³)				
	Cut	Fill	Cut	Fill	Adjusted Fill	Net	Cumulative
0 + 00.00	563.2	342.2					
0 + 50.00	213.5	213.6	719.2	−514.6	−584.8	134.4	134.4
1 + 00.00	123.5	343.3	312.0	−515.6	−586.0	−273.9	−139.6
1 + 50.00	654.6	111.0	720.5	−420.6	−478.0	242.5	102.9
2 + 00.00	973.1	762.4	1507.1	−808/7	−919.0	588.1	691.0
2 + 50.00	567.3	342.9	1426.3	−1023.4	−1163.0	263.3	954.3
3 + 00.00	451.6	190.4	943.4	−493.8	−561.1	382.3	1336.6

(Continued on next page)

As an example: between station $0 + 0.00$ and $0 + 50.00$, volume of fill $= \frac{1}{2}\,(342.2 + 213.6) \times 50 = 13{,}895$ ft³ $= 514.6$ yd³

Adjusted fill $= 514.6 \div 0.88 = 584.8$ yd³ (shown highlighted)

Net earthwork volume (sum of all adjusted fill and all cut volumes) $= 1{,}336.6$ yd³   **(A)**

602

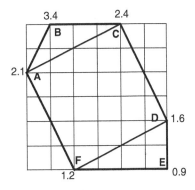

It helps to recognize that in the given geometry, the lines AF and CD are parallel and so are AC and FD, and that these lines are of the same length, thereby forming a square.

Thus, the volume can be found by adding together the volume for the shapes ABC, ACDF, and DEF.

Area of triangle ABC (base BC $= 150$ ft, height $= 100$ ft) is:

$$A_{ABC} = \frac{1}{2}\text{bh} = \frac{1}{2} \times 150 \times 100 = 7{,}500 \text{ ft}^2$$

Average depth of cut, $d_{ave} = \frac{1}{3}(2.1 + 3.4 + 2.4) = 2.63$ ft

Volume of cut in triangle ABC: $V_{ABC} = 7{,}500 \times 2.63 = 19{,}725$ ft³

Area of square ACDF (sides 223.6 ft) is:

$$A_{ACDF} = 223.6 \times 223.6 = 50{,}000 \text{ ft}^2$$

Average depth of cut, $d_{ave} = \frac{1}{4}(2.1 + 2.4 + 1.6 + 1.2) = 1.83$ ft

Volume of cut in square ACDF: $V_{\text{ACDF}} = 50{,}000 \times 1.83 = 91{,}500 \text{ ft}^3$

Area of triangle DEF (base EF = 200 ft, height DE = 100 ft) is:

$$A_{\text{DEF}} = \frac{1}{2}bh = \frac{1}{2} \times 200 \times 100 = 10{,}000 \text{ ft}^2$$

Average depth of cut, $d_{\text{ave}} = \frac{1}{3}(1.6 + 0.9 + 1.2) = 1.23 \text{ ft}$

Volume of cut in triangle DEF: $V_{\text{DEF}} = 10{,}000 \times 1.23 = 12{,}300 \text{ ft}^3$

Total volume = 123,525 ft³ = 4,575 yd³ (A)

A longer (but more general) solution would subdivide the region into triangles ABC, ACF, CFD, and DEF.

603

Rod reading at benchmark = 7.85. Therefore, height of instrument = 154.45 + 7.85 = 162.30 ft

Rod reading at A = 8.92 ft. Therefore, elevation at A = height of instrument − rod reading = 162.30 − 8.92 = 153.38 ft

Answer is (C)

604

Identify the edges of the limit of free haul by looking for two stations 300 ft apart which have equal cumulative yardage ordinates. Note that the mass diagram ordinates are equal (1,275 yd³) at stations 45 + 10 and 48 + 10.

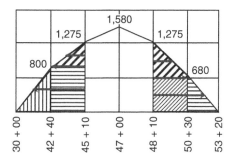

(Continued on next page)

The total overhaul is computed by taking moments of the mass diagram about the limits of free haul (stations 45 + 10 and 48 + 10) to the left and right, respectively, up to the balance points (stations 30 + 00 and 53 + 20, respectively). For each segment of the diagram, the overhaul is calculated as the product of the vertical ordinate difference (Δy) by the distance to the center of that interval. For example, the volume of 680 between stations 50 + 30 and 53 + 20 is triangular and the arm used extends from the limit of free haul (station 48 + 10) to the midpoint of the interval (50 + 30 to 53 + 20), which is at 51 + 75. These arms are shown with arrows in the figure here.

Overhaul = $475 \times 1.35 + 800 \times 8.90 + 595 \times 1.10 + 680 \times 3.65 = 10{,}898$ yd^3-sta

Cost of overhaul = $10{,}898 \times 3.7 = \$40{,}322$ **(B)**

605

Pursuing option B has: Initial cost = $200,000

Monthly savings = $17{,}000 - 8{,}000 = \$9{,}000$

Resale value (after 18 months) = $120,000

Monthly interest rate (nominal) = $10/12 = 0.833\%$

$$\left(\frac{A}{F}, 0.833\%, 18 \text{ periods}\right) = \frac{i}{(1+i)^n - 1} = \frac{0.00833}{1.00833^{18} - 1} = 0.057124$$

$$\left(\frac{A}{P}, 0.833\%, 18 \text{ periods}\right) = \frac{i(1+i)^n}{(1+i)^n - 1} = \frac{0.00833 \times 1.00833^{18}}{1.00833^{18} - 1} = 0.060057$$

Converting everything to annuities: Initial cost = $200{,}000 \times 0.060057 = \$12{,}011$

Monthly benefit = 9,000

Cost offset (resale) = $120{,}000 \times 0.051724 = \$6{,}207$

Benefit cost ratio: $\dfrac{B}{C} = \dfrac{B}{C - S} = \dfrac{9{,}000}{12{,}011 - 6{,}207} = 1.55$ **(B)**

606

At the rate of return (i), the present worth should be zero.

The $350k capital expenditure is a present value (P)	NEGATIVE
The $25k reduction in annual costs is an annuity (A)	POSITIVE
The $200k increase in salvage value is a future sum (F)	POSITIVE

Converting all of these to present worth, the net present worth can be written:

$$PW = -350 + 25\left(\frac{P}{A}, i, 20 \text{ yrs}\right) + 200\left(\frac{P}{F}, i, 20 \text{ yrs}\right) = 0$$

For $i = 5\%$, PW = 37k

For $i = 6\%$, PW = -0.9k

Therefore, the interest rate is a little less than 6%. Exact answer 5.97%

Answer is   (B)

607

Since the side-slopes are 3:2 and the horizontal projection of the side-flares are 30 ft, the depth of the channel is 20 ft and the inclined sides are $\sqrt{30^2 + 20^2} = 36.1$ ft

Total channel perimeter = 19 ft + 2 × 36.1 = 91.2 ft

Total surface area = 91.2 × 450 = 41,040 ft²

Cost of finishing compound (including waste) = 41,040 ft² ÷ 300 ft²/gal × 1.05 = 143.64 gal = 28.7 × 5-gal containers. Use 29 containers. Cost = 29 × $40/container = $1,160.00

Total cross sectional area = 91.2 ft × 8 in = 60.8 ft²

Total volume of concrete = 60.8 × 450 = 27,360 ft³ = 1,013.33 yd³

Cost of concrete (including waste) = 98 × 1013.33 × 1.05 = $104,272

Total cost of materials = $105,432   (D)

608

Since the side slope is 1:3, a vertical dimension of 12 ft corresponds to a horizontal dimension of 36 ft. Therefore, the plan dimensions of the top of the landfill are 3,928 ft × 2,428 ft. This gives a plan area = 9,537,184 ft^2

The volume of the cover soil can be approximated as top area × depth = 9.5372 × 10^6 × 1.5 = 1.431 × 10^7 ft^3 = 530,000 yd^3 (A)

Note: A more precise solution can be calculated by the formula of a pyramidal frustum, which shows that the approximate method yields only 0.5% error. This is because the thickness of the soil in question (18 in) is so small compared to the plan dimensions.

Frustum of pyramid: $V = \dfrac{h}{3}\left(A_1 + A_2 + \sqrt{A_1 A_2}\right) \approx \dfrac{h}{2}\left(A_1 + A_2\right)$

609

Area of walls (single layer): $A = 2 \times (180 + 200) \times 12 - 8 \times 6 \times 10 = 8{,}640$ ft^2

Fully burdened cost for labor:

$4 \times 40 + 2 \times 25 + 1 \times 50 = \$260$ per crew hour = \$37.14 per labor hour

Credit for 5/8 in GWB (1 layer):   Labor = 8,640 ft^2 ÷ 960 ft^2/LH × \$37.14/LH
        = \$334.27
        Materials = 8,640 ft^2 × \$0.255/ft^2
        = \$2,203.20 + 10% waste = \$2,423.52

½ in. GWB (2 layers):   Labor = 17,280 ft^2 ÷ 960 ft^2/LH × \$37.14/LH
        = \$668.52
        Materials = 17,280 ft^2 × \$0.285/ft^2
        = \$4,924.80 + 10% waste = \$5,417.28

Insulation (1 layer):   Labor = 8,640 ft^2 ÷ 1,920 ft^2/LH × \$37.14/LH
        = \$167.13
        Materials = 8,640 ft^2 × \$0.45/ft^2
        = \$3,888.00 + 10% waste = \$4,276.80

Cost differential (new SOW − original SOW) = − 2,757.79 + \$6,085.80 + 4,443.93 = \$7,771.94

After applying 1.10 and 1.05 factors for overhead and profit, the net increase in the change order = \$9,831.50 (C)

610

By employing one extra crew unit, work will be completed in another $90 - 15 = 75$ days

For these 75 days, an extra crew unit will cost an additional $= 3,200 \times 75 = \$240,000$

Early completion bonus $= 18,000 \times 15 = \$270,000$

Net bonus $= \$30,000$ (acceptable)

By employing two extra crew units, work will be completed in $90 - 21 = 69$ days

For these 69 days, two extra crew units will cost an additional $= 6,000 \times 69 = \$414,000$

Early completion bonus $= 18,000 \times 21 = \$378,000$

Net bonus $= -\$36,000$ (unacceptable)

By employing three extra crew units, work will be completed in $90 - 31 = 59$ days

For these 59 days, three extra crew units will cost an additional $= 9,000 \times 59 = \$531,000$

Early completion bonus $= 18,000 \times 31 = \$558,000$

Net bonus $= \$27,000$ (acceptable, but less than net profit with one additional crew unit) (**A**)

611

Dump truck: Ideal travel time $= 8$ mi round trip $\div 30$ mph $= 0.267$ hr $= 960$ sec

Ideal transfer time $= 40$ sec per 3 yd^3 $= 213$ sec per 16 yd^3 load

Ideal dumping time $= 30$ sec per load

Cycle time $= 1,203$ sec

Dump truck productivity (ideal) $= 16$ yd^3 per 1,203 seconds $= 0.0133$ yd^3/sec (loose soil)

Dump truck productivity (actual) $= 0.7 \times 0.8 \times 0.0133 = 0.00745$ yd^3/sec (loose soil)

Rate of power shovel excavation $= 10$ yd^3/min (bank) $= 11$ yd^3/min (loose soil)

For each 3 yd^3 (loose soil) load of the power shovel, excavation time $= 3 \div 11$ min $= 16.4$ sec

Cycle time of power shovel $=$ excavation time $+$ transfer time $+$ dump time $= 16.4 + 40 + 30 = 86.4$ sec

Power shovel production (loose soil) $= 3$ yd^3/86.4 sec $= 0.03472$ yd^3/sec

(*Continued on next page*)

Number of dump trucks needed to balance the power shovel production = 0.03472 ÷ 0.00745 = 4.66

Use five trucks. (C)

612

The time needed by each excavator can be calculated by dividing 18,000 yd³ by the production rate (yd³/hr). This will yield the "time to complete" in hours, which can then be multiplied by the variable cost rate ($/hr) to obtain the variable cost. The fixed cost can then be added in.

Total cost = Fixed cost + 18,000 yd³ ÷ Production (yd³/hr) × Variable cost ($/hr)

The total cost for types 1–4 are $15,094, $13,214, $19,750, and $14,700, respectively.

Answer is (B)

613

Even though the problem is three-dimensional, we can take advantage of the symmetry in the problem. All cables carry equal load. The length of each cable is given by:

$$L = \sqrt{5^2 + 8^2 + 2^2} = 9.644 \text{ ft}$$

The vertical component of each cable tension can therefore be written:

$$T_y = T\frac{8}{9.644} = 0.83T$$

Therefore, the equation of vertical equilibrium is:

$$4T_y - 900 = 0 \Rightarrow 4 \times 0.83T = 900 \Rightarrow T = 271 \text{ lb} \tag{A}$$

614

Design capacity = 40 tons

Since $FS = 6$, ultimate capacity = $40 \times 6 = 240$ tons = 480,000 lb

Pile hammer energy $WH = 50,000$ ft-lb = 600,000 in-lb

$$Q_{ult} = \frac{WH}{S + 1.0} \Rightarrow S = \frac{WH}{Q_{ult}} - 1 = \frac{600,000}{480,000} - 1 = 0.25$$

$$S = 0.25\frac{\text{in}}{\text{blow}} \Rightarrow 4\frac{\text{blows}}{\text{in}} = 48\frac{\text{blows}}{\text{ft}} \tag{D}$$

615

The southwest corner of the mat is furthest (radially) from the wellpoint, at a distance:

$$r = \sqrt{190^2 + 240^2} = 306.1 \text{ ft}$$

Water table elevation at this point needs to be 325.65 − 5.0 = 320.65 ft. Drawdown = 325.8 − 320.65 = 5.15 ft

At the southwest corner of the site, the radial distance is $r = \sqrt{240^2 + 300^2} = 384.2$ ft). Here, the drawdown is 325.8 − 324.0 = 1.8 ft.

Applying the steady-state equation to these two points:

$$Q = \frac{2\pi T(s_2 - s_1)}{\ln(r_2/r_1)} = \frac{2\pi \times 250 \times (5.15 - 1.8)}{\ln(384.2/306.1)} = 23,156 \frac{\text{ft}^3}{\text{hr}} = 6.43 \frac{\text{ft}^3}{\text{sec}} = 2,887 \text{ gpm} \quad \textbf{(D)}$$

616

Cycle time of the excavator = 45 sec + 4 min + 45 sec = 5 min 30 sec = 5.5 min

Ideal volume (loose measure) turnover = 3.0 yd^3/5.5 min = 0.545 yd^3/min

Actual volume turnover = 0.9 × 0.545 = 0.491 yd^3/min

Truck production = 15 yd^3/90 min = 0.167 yd^3/min/truck

The number of trucks needed to balance the production of the excavator = 0.491 ÷ 0.167 = 2.95

Use three trucks. **(B)**

Note: Since this is a question of balance, the total size of the job does not matter.

617

Material delivery = 950 yd^3/hr (bank measure), which is equivalent to 950 × 0.85 = 807.5 yd^3/hr (compacted) after shrinkage is accounted for.

Roller covers ground at 3 mph × 8 ft = 126,720 ft^2/hr.

Since a 6-in (0.5 ft)-thick layer gets compacted in four passes, each pass compacts the equivalent of 0.125 ft, which means it compacts 126,720 × 0.125 = 15,840 ft^3 (586.7 yd^3) of soil per pass.

(Continued on next page)

This is ideal capacity. Working at 80% efficiency, the roller compacts $0.8 \times 586.7 = 469.3$ yd^3/hr

No. of rollers required $= 807.5/469.3 = 1.72$. Therefore, two rollers are needed to handle the delivery of the material. (B)

618

Since the outriggers are simply resting on the soil (not anchored), the limiting condition is when the far side outrigger legs have zero reaction. For this condition, the inside legs (pair on the right) carry the total load of 40 tons. Taking moment about the inside leg:

$40(X - 10) = 25 \times 10$

Solving for X: maximum offset $X = 16.25$ ft (B)

619

Top of pile cap at depth $= 3$ ft

Bottom of pile cap at depth $= 5$ ft

Top of piles at depth $= 4$ ft

Bottom of piles at depth of rock layer $= 34$ ft

Pile length required $= 30$ ft (D)

620

The network diagram constructed from the information in the table is shown in the following figure. Each activity is shown as a block which shows (starting with upper left corner, going clockwise—ES (early start), EF (early finish), LF (late finish), LS (late start).

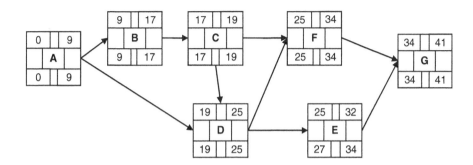

Project duration is the finish time for the terminal activity = 41 days (C)

621

According to the PERT model, mean and standard deviation are given by:

$$\mu = \frac{a + 4b + c}{6} \qquad \sigma = \frac{c - a}{6}$$

Completing the table with the mean (μ) and standard deviation (σ) for tasks A, C, E, and F

Activity	Duration (weeks)		
	μ	σ	σ^2
A	4	0.33	0.11
C	6.83	0.50	0.25
E	6	0.33	0.11
F	8.83	0.50	0.25

Therefore, for the activity sequence ACEF, the mean is the sum of the means and variance is the sum of the variances: $\mu = 25.66$, $\sigma^2 = 0.72 \rightarrow \sigma = 0.85$

For X = 25, the standard normal variable: $Z = \dfrac{X - \mu}{\sigma} = \dfrac{25 - 25.66}{0.85} = -0.776$

Probability of completion time being less than 25 weeks is calculated as:

$P(Z \le -0.776) = 1 - P(Z \le +0.776) = 1 - 0.7811 = 0.2189 = 22\%$ (D)

622

Forward pass through network results in: $ES_A = 0$, $EF_A = 3$, $ES_C = 3$, $ES_D = 6$

Free float of activity A = min ES of all successors (C & D) − EF of A = $3 - 3 = 0$ **(A)**

623

In the original plan (clear rectangles), E ends at 5 weeks 2 days, G begins at 5 weeks. Therefore, stage G can start without completion of stage E. Option A is incorrect.

Stage F is on schedule. Option B is incorrect.

Completion levels are (approx.): A (75%), B (85%), C (85%), D (25%), E (25%), F (25%), and G (0%). At current time (5 weeks, 2 days), they should have been A (100%), B (90%), C (100%), D (75%), E (45%), F (25%), and G (0%). Stages A, B, C, and D are behind schedule. Option C is correct.   **(C)**

624

Current bid price can be calculated by applying inflation (3.2%) factor to the price on file (2½ years ago):

Cost per dump truck load = $525 \times (1 + 0.032)^{2.5} = 568$

Excavated material 323,000 ft³ will swell to $323,000 \times 1.25 = 403,750$ ft³ = 14,954 yd³

This will occupy $14,954 \div 26 = 575$ truckloads

Bid price should be $575 \times 568 = \$326,600$ for the 14,954 yd³ of earth moved   **(C)**

625

Maximum moment, $M = \dfrac{PL}{3} = \dfrac{800 \times 60}{3} = 16,000$ lb · in

Section modulus of rectangular beam section, $S = \dfrac{1}{6}bh^2 = \dfrac{6^3}{6} = 36$ in³

Maximum bending stress, $\sigma = \dfrac{M}{S} = \dfrac{16,000}{36} = 444.44$ psi

This is the modulus of rupture, which is correlated to f_c' according to:

$$f_r = 7.5\sqrt{f_c'}$$

Therefore, $f_c' = 3,511$ psi $\hfill$ (D)

626

Split cylinder test tensile strength: $f_{ct} = \dfrac{2P}{\pi DL} = \dfrac{2 \times 55,000}{\pi \times 6 \times 12} = 486$ psi

$$f_{ct} = 6.7\sqrt{f_c'}$$

Therefore, $f_c' = 5,262$ psi $\hfill$ (B)

627

The weld appears on both sides (arrow side and far side). Therefore, I is incorrect.

The fillet weld symbol (triangle) in figure III is flipped (vertical edge on the right). Therefore, it is incorrect.

The weld does not appear to be an all-around weld. Therefore, IV is incorrect.

Answer is $\hfill$ (D)

628

| Cement: | Weight = 160 lb |
| | Volume = $160/(3.15 \times 62.4) = 0.814$ ft^3 |

Sand (wet):	Weight = 290 lb
Sand (SSD):	Weight = $100.7/105 \times 290 = 278.12$ lb
	Volume = $278.12 \div (2.7 \times 62.4) = 1.651$ ft^3

Free water in sand = $4.3/105 \times 290 = 11.88$ lb

Coarse aggr. (wet):	Weight = 420 lb
Coarse aggr. (SSD):	Weight = $100.5/103 \times 420 = 409.81$ lb
	Volume = $409.81 \div (2.6 \times 62.4) = 2.526$ ft^3

Free water in coarse = $2.5/103 \times 420 = 10.19$ lb

(*Continued on next page*)

Total water: Weight $= 56 + 11.88 + 10.19 = 78.07$ lb

Volume $= 78.07 \div 62.4 = 1.251$ ft^3

Total volume of these components $= 0.814 + 1.651 + 2.526 + 1.251 = 6.242$ ft^3

This volume represents 97% of total volume (since air $= 3\%$).

So, total volume $= 6.242/0.97 = 6.435$ ft^3

Unit weight $=$ Total weight/total volume $= 926$ lb$/6.435$ ft$^3 = 143.9$ lb/ft^3 (C)

629

From the curve, corresponding to a strength of 3,500 psi, we obtain $TTF = 1,800°$F-hr.

Therefore, the time interval for sufficient maturity is given by:

$$\Delta t = \frac{TTF}{T - T_o} = \frac{1,800}{70 - 30} = 45 \text{ hr}$$

Contractor has to wait at least 45 hr after the concrete is placed before removing the forms. (C)

630

Probability of failure, $p = 1 - 0.90 = 0.10$

No of failures $= 0$

Based on the Binomial distribution, probability of R defects in a sample size N is

$$P = C\left(\begin{array}{c} N \\ R \end{array} \right) p^R (1 - p)^{N-R}$$

Setting this equal to $1 - 0.80 = 0.20$, and $R = 0$, we get:

$$1 \times 0.1°0.9^N = 1 - 0.8 = 0.2 \Rightarrow N = 15,3$$

Therefore, minimum sample size of defect-free cylinders $= 16$, to achieve 80% confidence. (A)

631

At a height $H = 14$ ft above the bottom of the footing, the width of the Rankine zone is

$$L_R = 14\cot(45 + \phi/2) = 8.1 \text{ ft}$$

At a depth of 10 ft, the total overburden pressure $= 120 \times 10 = 1{,}200$ psf

Active earth pressure coefficient: $K_a = \dfrac{1 - \sin\phi}{1 + \sin\phi} = 0.333$

Lateral pressure at a depth of 10 ft $= K_a \gamma H = 0.333 \times 120 \times 10 = 400$ psf

Lateral force on the blanket at depth of 10 ft $= 3 \text{ ft} \times 400 \text{ psf} = 1{,}200 \text{ lb/ft}$

Therefore, the effective length of reinforcing strip required is given by

$$L_E = \frac{1{,}200}{2 \times \tan 15 \times 1{,}200} = 1.87 \text{ ft}$$

Therefore, total length of reinforcing strip at depth $D = 10$ ft is
$L = 8.1 + 1.9 = 10$ ft   (B)

632

From ACI-347, for wall forms, for rate of pour $= 4$ ft/hr and concrete temp $= 60°F$, base value (from tables, or equation) of the lateral pressure $= 750$ psf

For type I cement with retarder, the adjustment factor $= 1.2$

For lightweight concrete ($\gamma = 135$ pcf), the adjustment factor $= 0.5\left(1 + \dfrac{135}{145}\right) = 0.966$

Therefore, the adjusted lateral pressure $= 750 \times 1.2 \times 0.966 = 869.4$ psf   (A)

633

According to OSHA 3146: "Anchorages used to attach personal fall arrest systems shall be independent of any anchorage being used to support or suspend platforms and must be capable of supporting at least 5,000 lbs (22.2 kN) per person attached."   (A)

634

The total load of $D + 0.3D + 0.1D$ from slab no. 5 is distributed equally through interconnected slabs 2, 3, and 4. Thus, each of these three floors carries $1.4D/3 = 0.47D$.

The total load transmitted to the reshores below level $4 = D + 0.3D + 0.1D + D + 0.05D = 2.45D$, of which slab 4 carries $1.47D$. Therefore, the load transmitted down to the reshores below level $4 = 2.45D - 1.47D = 0.98D$.

Answer is   (D)

635

According to the *Standard Practice for Bracing Masonry Walls during Construction,* whenever a masonry wall is constructed, a limited access zone is to be established, whose width shall be wall height plus 4 ft.   (C)

636

According to ASCE 37, the minimum nominal value of the horizontal construction load is taken as the largest of:

1. For wheeled vehicles transporting materials is the larger of 0.2 times the fully loaded weight (for a single vehicle) or 0.1 times the fully loaded weight (for multiple vehicles).
2. 50 lb per person at the level of the platform.
3. 2% of the total dead load.
4. The calculated horizontal reaction.

These estimates are, respectively:

1. larger of $0.2 \times 600 = 120$ lb or $0.1 \times 1800 = 180$ lb
2. 100 lb
3. 300 lb
4. Not provided

Answer is   (D)

637

The Rankine failure plane is oriented at $\alpha = 45 + \phi/2$ as shown in the figure. From the back face of the wall, the distance $X = 5 + 16/\tan 62 = 13.5$ ft. Any load on the backfill must be placed beyond this surface.

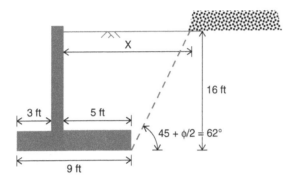

Answer is (C)

638

According to OSHA 29CFR 1926, the permissible exposure thresholds corresponding to noise levels 90, 95, and 100 dB are:

90 dB—8 hr

95 dB—4 hr

100 dB—2 hr

The noise exposure factor is calculated as:

$$F(e) = \frac{3}{8} + \frac{1}{4} + \frac{1}{2} = 1.125 \qquad \text{(D)}$$

639

Total number of hours worked by all employees $= 400 \times 1{,}970 = 788{,}000$

Total recordable cases $= 13 + 18 = 31$

OSHA recordable case rate $= 31 \times 200{,}000/788{,}000 = 7.87$ (B)

640

According to MUTCD, the location of the ROAD WORK AHEAD sign from the center of the work zone = 1,250/2 + distances A, B, and C = 625 + 1,000 + 1,500 + 2,640 = 5,765 ft
(B)

Table 6H-3. Meaning of Letter Codes on Typical Application Diagrams

Road Type	Distance Between Signs**		
	A	B	C
Urban (low speed*)	30 (100)	30 (100)	30 (100)
Urban (high speed*)	100 (350)	100 (350)	100 (350)
Rural	150 (500)	150 (500)	150 (500)
Expressway/Freeway	300 (1,000)	450 (1,500)	800 (2,640)

Answer Key for Construction Depth Exam

601	A
602	A
603	C
604	B
605	B
606	B
607	D
608	A
609	C
610	A

611	C
612	B
613	A
614	D
615	D
616	B
617	B
618	B
619	D
620	C

621	D
622	A
623	C
624	C
625	D
626	B
627	D
628	C
629	C
630	A

631	B
632	A
633	A
634	D
635	C
636	D
637	C
638	D
639	B
640	B